MATERIALS SCIENCE AND TECHNOLOGIES

FOCUS ON ACOUSTIC EMISSION RESEARCH

MATERIALS SCIENCE AND TECHNOLOGIES

Additional books in this series can be found on Nova's website under the Series tab.

Additional e-books in this series can be found on Nova's website under the eBook tab.

MATERIALS SCIENCE AND TECHNOLOGIES

FOCUS ON ACOUSTIC EMISSION RESEARCH

CLAUDIA BARILE AND
GIOVANNI PAPPALETTERA
EDITORS

nova
science publishers
New York

Library of Congress Cataloging-in-Publication Data

ISBN: 978-1-53611-848-3

Published by Nova Science Publishers, Inc. † New York

CONTENTS

Preface		**vii**
Chapter 1	The Identification of Damage Mechanisms with Acoustic Emission Monitoring: Interests and Limitations *Nathalie Godin, Pascal Reynaud, Mohamed R'Mili and Gilbert Fantozzi*	**1**
Chapter 2	Fatigue Damage Propagation in AISI 430 F/1: A Monitoring Approach Based on Acoustic Emission and Infrared Thermography *Claudia Barile, Caterina Casavola, Giovanni Pappalettera and Carmine Pappalettere*	**53**
Chapter 3	The Application of Acoustic Emission Technique in the Delamination of Laminated Composites *Mohamad Fotouhi, Milad Saeedifar, Jalal Yousefi and Sakineh Fotouhi*	**73**
Chapter 4	Acoustic Emission Research on Plant Fibre Composites *Carlo Santulli*	**113**
Chapter 5	Acoustic Emission Monitoring of In-Service Large-Sized Structures *Gianni Niccolini, Giuseppe Lacidogna and Alberto Carpinteri*	**133**

Editors' Contact Information 151

Index 153

PREFACE

The application of acoustic emission technique has greatly grown in the last decades in several industries areas. This appears evident by checking the increasing number of papers, documents and researches that have progressively enriched scientific literature. The basic principle of this technique refers to the possibility of detecting stress and/or pressure waves produced by dynamic process inside the materials. One of the first applications of acoustic emission detection to monitor oncoming dangerous events probably occurred in seismology in which elastic waves were associate to earthquakes. Similarly, acoustic emission generated by timber under heavy loads was used to indicate the imminent failure of wooden structures. Finally, sounds emitted by rocks were used to detect the onset of landslides. Pioneers works that systematically implement the acoustic emission analysis were done by Kaiser and then Tatro and Schofield. Kaiser was the first to use electronic instrumentation to detect audible sounds derived from deformation of metals. Successively Schofield and Tatro search out the acoustic emission sources in dislocation motion associating to plastic deformation rather than being totally due to grain boundary sliding. For these reasons, all researchers approaching this theme must to thank them and keep in mind their methods as examples.

The aim of this volume is to underline the benefits derived by this technique if applied for scientific and engineering purposes and how the "sounds" emitted by materials and structure can be used to evaluate the structural integrity of components. Differently from other acoustic emission books, this volume intends to illustrate the technique by showing benefits and limits but above all to describe current applications and the most promising results. Its intention is to improve the technical knowledge and to enrich the scientific community. The first section introduces the technique and defines the main signals connected with the damage mechanism for different kind of materials. Then the limits of the technique are described in order to show the need of a quantitative, AE based on, modelling works. Each of the following sections describes a specific material or components application: metals, laminated composites, plant fiber composite, large scale structures.

On a personal note, this has been my first involvement as editor. I would like to express my honest appreciation and gratitude to all chapter authors of this book and to my co-editor. They supported me in this important challenge sharing their experience together with their own specialization. Their sincere involvement to share their expertise makes them great researchers as well as great colleagues. They have been extraordinarily patient with all my requests, working hardly and patiently. And finally, more personally, I would like to thank my family for taking care of me in everyday life. You are my shoulder and my strength!

Claudia Barile
Bari, Italy
January 2017

In: Focus on Acoustic Emission Research ISBN: 978-1-53611-848-3
Editors: C. Barile and G. Pappalettera

Chapter 1

THE IDENTIFICATION OF DAMAGE MECHANISMS WITH ACOUSTIC EMISSION MONITORING: INTERESTS AND LIMITATIONS

Nathalie Godin**, Pascal Reynaud, Mohamed R'Mili and Gilbert Fantozzi
Mateis, Insa de Lyon, University of Lyon, Villeurbanne, France

ABSTRACT

Damage monitoring of structural materials is a key point for the control of durability and reliability in service. Beyond the characterization of damage, the current challenge is to predict the lifetime of the structures in service. To achieve this goal, a quantification of damage is needed but mainly an identification of the various damage mechanisms, for which acoustic emission (AE) is a well-adapted technique. This chapter focuses on the identification of acoustic signature of several damage mechanisms obtained with a clustering method (k-means, PCA, k-nearest neighbors, Kohonen map, Random forest….). Studies are conducted on several kinds of materials such as fibers composite, metastable austenitic stainless with strain-induced martensitic transformation at room temperature, refractories with very high content zirconia. The objective of this paper is to present an approach to damage

* Corresponding Author: nathalie.godin@insa-lyon.fr.

identification based on signal analysis by supervised and unsupervised classification. Therefore, with a clustering method, AE signals, which have similar shapes or similar features, can be grouped together into a cluster. Afterwards, each cluster can be linked with a main damage. In this way, a careful analysis of acoustic emission signals can lead to the discrimination of the different damage mechanisms occurring in a material. It is a possible solution for the identification of damage during service with a view to component lifetime control. On the other hand, the limitation of this approach is also investigated. Ground truth data coming from the synthetic dataset are also used in order to validate and to show the main limitations of this approach (pertinent descriptors, minor damage mechanism, etc.). Moreover, the identification of the acoustic signature is done for given testing conditions and sample geometry. Sizes effects are not taken into account and are investigated with modelling work.

Keywords: damage identification, pattern recognition, lifetime

1. INTRODUCTION

A challenge concerning materials and structures is their residual lifetime prediction when subjected to a wide variety of environmental and mechanical loading conditions, which can initiate damage and lead to failure. The key issue in predicting the lifetime of many different parts is obviously to use the results given by the continuous health monitoring. It appears necessary to understand the mechanisms of damage initiation, evolution and criticality and so, to identify the relevant precursors of failure. Sound identification of damage mechanisms and kinetics is an important issue with a view to predict remaining lifetime. For the real time in-situ monitoring of these phenomena, the acoustic emission is of particular interest [1, 2]. Indeed, acoustic emission, which corresponds to the part of energy generated by the material during the damage processes, is directly related to the damage mechanisms and so, can give pertinent information about the damage initiation and development. Therefore, acoustic emission testing is a promising technique for the monitoring of structural integrity owing to its ability to detect the mechanical energy released by flaw evolution with a high sensitivity. The sources of AE are varied: plastic deformation localized in metal alloys [3], initiation and propagation of cracks [4, 5], ruptures of fibers [6] or particles [7], interfacial decohesion, and phase transformation such as martensitic transformation (MT) [3]. Therefore, this technique potentially allows not only the location of the source of the emission, but also the determination of its nature. However, the stress waves resulting from the microstructural changes depend on the propagation conditions such as the ultrasonic attenuation in a heterogeneous medium, and on boundary surface interactions. Therefore, the signal delivered by the sensor is a strongly modified representation of the original source. Nevertheless, it is realistic to consider that this signal contains some features representative of the source in such a manner that correlation exists between the damage mechanisms and the magnitude of the various AE parameters. Consequently, each signal can be considered as the acoustic signature of the different damage modes. However, the main problem associated with data analysis is the discrimination between the different acoustic emission sources. The conventional analysis of AE data is based on the parameters extracted from the AE signal. Numerous studies have attempted to differentiate the source mechanisms of AE signals by using a single parameter. Most of them took the amplitude as the principal parameter of AE signals [8-16]. Some other authors have chosen to work on the frequency [17, 18], the peak frequency [19], and the "rise angle" (defined by the ratio amplitude/rise time) [17, 20]. For instance, Barré [15] identified four areas of amplitude, corresponding to the matrix cracking, debonding, loosening and breaking of fibers for fibers reinforced composites. Other authors have used a multivariable analysis using two or more descriptors [21-24]. Kostopoulos [25-27], Godin [28, 29], Moevus [30, 31] have identified different classes of AE signals which are attributed to damage modes in oxide/oxide composite, glass/polyester and SiC_f/(Si-B-C) composites. In [31], the AE data are analyzed using an unsupervised multi-variable clustering technique. It is shown that several types of signals can be distinguished, and that the obtained clusters are consistent with the expected damage modes in 3D SiC_f /SiC composites. For the corrosion processes, in 1996, Barton et al. [32] developed an artificial neural network (ANN) to identify the onset and classify the type of localized corrosion from electrochemical noise (ECN) spectra. Van Dijck and Hulle [33] used a hybrid filter-wrapper genetic algorithm and a naive Bayes classifier to identify the absence of corrosion, uniform corrosion, pitting and stress corrosion cracking.

AE based identification of damage modes is not straightforward for two reasons. On the one hand, the AE signals have complex shapes that must be characterized by multiple pertinent descriptors. On the other hand, the acoustic signatures of damage modes are not known *a priori* and they are scattered.

The objective of the cluster analysis [34-36] is to separate a set of data into several classes that reflect the internal structure of the data. Indeed, cluster analysis is an important tool for investigating and interpreting data.

In order to associate each waveform with a specific mechanism of damage, we need a method to define the similarity or the differences between the signals of AE. A vector with n parameters or n descriptors characterizes the signals. Therefore, the main problem during clustering would be that we do not know a priori what the real cluster structure is (number and composition) although the damage mechanisms are known.

The approach involves two main steps:

(1) AE signals which have similar characteristics are grouped using a clustering method in order to identify acoustic signatures of the damage mechanisms;
(2) Then a supervised classification is conducted, using a library of labelled pre-classified signals that can be used in real time.

In this approach, the acoustic emission is one additional tool for a better understanding of the damage mechanisms. However, the acoustic emission, once it is well characterized and controlled, can become a predictive tool [37-39]. Once the acoustic signature of the various mechanisms sources is identified, a library of labelled signals is thus available in order to do a supervised classification. That will allow in real time to identify the type of damage that occurs and to have an idea of its severity.

The first part of this chapter is dedicated to the identification of the acoustic signature of damage's mechanisms for several kinds of materials. In the second part, the limits of these approaches are presented in order to show the need of a quantitative AE based on modelling works.

2. Identification Based on Clustering Methods

Acoustic emission signals can reasonably be considered as a representation of their sources and then the acoustic signature of damage mechanisms can be determined. In these studies, the following assumptions are made:

(1) Source mechanisms with different levels of released energy are associated with different waveforms.

(2) Signal energy is related to the energy released by the source mechanism. This parameter is thus very important for the discrimination of sources. It depends in particular on the source size and the stored elastic energy.
(3) Signals are affected by their propagation. The recorded signals depend jointly on mechanical sources, the material properties of the propagation medium and the sensors features.

Nevertheless, in permanent set-up conditions, similarities exist between AE signals originating from similar events.

A typical pattern clustering involves the following steps (Figure 1):

- representation of the pattern vectors (feature selection/extraction procedure),
- definition of a similarity measure appropriate to the data domain,
- clustering (or grouping of similar pattern vectors),
- cluster validity analysis (using a specific criterion of optimality),
- labeling the clusters (identifying the AE sources).

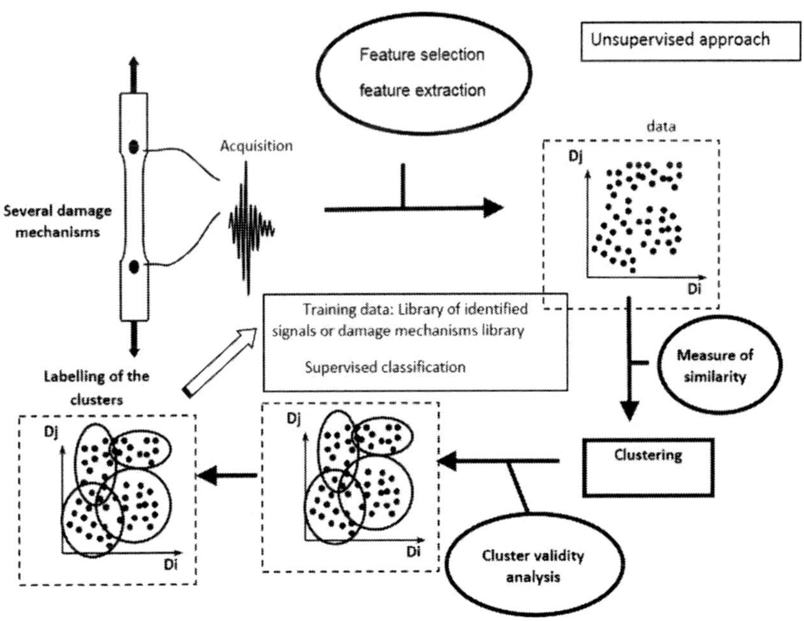

Figure 1. Schematic diagram of the pattern recognition method proposed for the analysis of the AE data.

In this part, we are going to summarize the procedure developed by several authors [3, 28, 31, 40-45].

2.1. Descriptors of Acoustic Emission Signal

2.1.1. Location of the AE Signal

The location of AE sources is determined from the difference in arrival times at sensors. During the uniaxial tensile test, AE is usually monitored by using a two-channel acquisition system. Since two sensors are used, location is determined with respect to the specimen longitudinal axis. So, only the signals coming from the working length of the specimens are analyzed.

The AE wave velocity is calibrated before the test, according to a pencil lead break procedure: several breaks are performed on the specimen at several locations x between the two sensors [46]. The difference in time of arrival $\Delta t(x)$ between the two sensors is calculated by using the first peak of each signal. The velocity Ce of an extensional wave in a thin plate is proportional to the square root of the elastic modulus E of the material. Since E decreases as damage occurs in the material, it is important to take into account the evolution of Ce during the mechanical test in order to better evaluate the location of the AE sources [47].

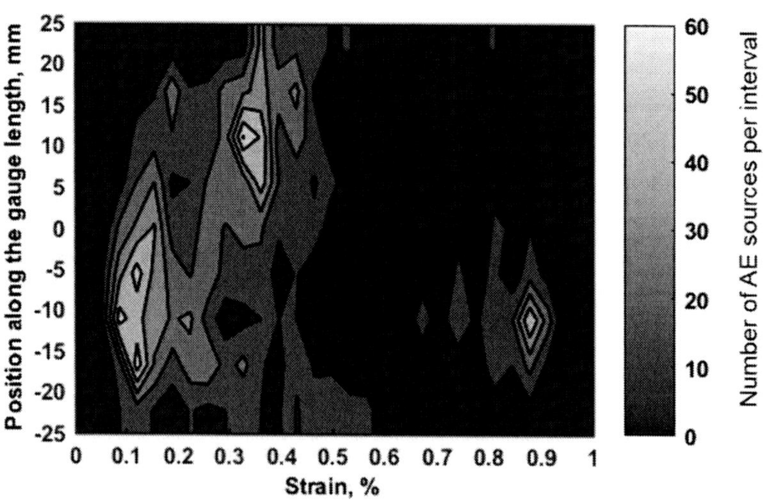

Figure 2. Location of the AE signal along the gauge length versus strain during a tensile test on CMC composite.

For example, at the end of the tensile test, the velocity on SiC_f /SiC composite is found to be equal to 6480 m/s, instead of 10000 m/s in the undamaged state. The decrease in wave velocity is thus not negligible. Figure 2 shows the position of AE signals recorded during tensile test on Ceramic Matrix Composite versus strain. This figure shows the evolution of linear density of acoustic events during monotonous loading. It can be noticed that the acoustic emission is very significant along the gauge length for strain level values between 0.10 and 0.30%. Just prior to final rupture, AE activity increases significantly in the rupture zone.

2.1.2. Definition of the Energy

It is generally accepted that the energy of an AE signal includes the energy released by the source at crack initiation. Various parameters affect recorded energy: distance of wave propagation, energy attenuation due to damage, coupling between sensor and material surface and sensor frequency response. Wave theory states that the energy of an acoustic wave decreases exponentially with the increase of the propagation distance. Therefore, the following equation is proposed to describe the energy of recorded AE signals (for instance, at sensor 1) received from the source n [38, 39]:

$$E_1(n) = E_s(n).A_1.e^{-B(L+x(n))} \tag{1}$$

$E_s(n)$ is the energy released at source n in the form of elastic waves. Due to differences in coupling between sensor and material surface or in sensor frequency response, for a source located at an equal distance, the sensors may record significantly different amounts of energy. Thus, A_i is the proportion of source energy that is recorded by sensor i. It is a constant characteristic of sensor. $L + x\,(n)$ is the distance of propagation from source n to sensor 1 (2L being the distance between sensors). The attenuation coefficient B is related to the propagation medium, which is subjected to changes due to damage evolution. Similarly, AE signal energy received at sensor 2 is expressed as:

$$E_2(n) = E_s(n).A_2.e^{-B(L-x(n))} \tag{2}$$

The source energy is then defined as the square root of the product of the amounts of energy received at both sensors for each source:

$$K(n) = \sqrt{E_1(n) \times E_2(n)} = A(n).e^{-B.L} \qquad (3)$$

Where A(n) is directly related to the energy of source n and B is related to the propagation medium but supposed to be independent of the source energy.

2.1.3. Pertinent Descriptors

AE signals (Figure 3) are described using several descriptors; amplitude, duration, energy, number of counts, rise time, counts at peak, frequency, etc. Table 1 and Table 2 summarize the descriptors introduced by Moevus [31] and Morizet [45]. In [31], the first nine (like the rise time RT or the amplitude A) are calculated by the acquisition system, where the others are ratios of the first ones. In Morizet et al. (2016) each waveform is turned into a compact representation through a set of 30 features, in time, frequency and wavelet domains (Table 2). Thus, an in-depth work has been realized to preprocess the corresponding waveforms, including wavelet denoising. Besides common features such as amplitude, duration, energy, rise time, partial powers or peak frequency, other features are derived from speech recognition and sound description studies. Wavelet features are actually a specific set of features using wavelet packet energy. The energy percentage of the terminal nodes of the wavelet packet tree is computed, leading to 23 = 8 wavelet packet energy features. Some authors [48], use the Hilbert-Huang transform for the extraction of a new relevant damage descriptor on non-stationary EA signals.

Particular care is taken to optimize the description of the AE signals, because the quality of the clustering result depends on the pertinence of the chosen descriptors for the data. Each descriptor X can be considered as a vector with n coordinates x_i, n is the number of AE signals. The values x_i of each descriptor are standardized as follows:

$$\overline{x}_i = \frac{x_i - \mu}{\sigma} \qquad (4)$$

Where $\overline{x}_i$ is the standardized value of x_i, μ is the average value of x_i over all AE signals processed, and σ is the standard deviation. Several descriptors, such as duration or energy, exhibit exponential distributions. It is therefore necessary to reduce their definition space by using their logarithmic value, to obtain comparable scales between all the AE descriptors after standardization.

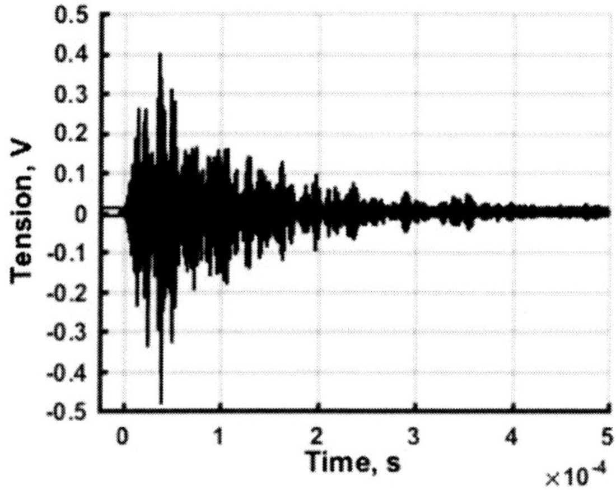

Figure 3. A representative acoustic emission signal.

Table 1. Descriptors of AE signals, the first 9 (like the rise time or the amplitude) are calculated by the acquisition system, the others are ratios of the first ones (Rise Angle: Amplitude/Rise Time)

Descriptor	Symbol	Unit
Rise Time	RT	μs
Counts	C	-
Duration	D	μs
Amplitude	A	dB
Average Frequency	AF	kHz
Counts to peak	CP	-
Decay Frequency	DF	kHz
Rise Frequency	RF	kHz
Energy	E	attoJ
Rise Time/Duration	RT/D	-
Duration/ Amplitude	D/A	μs/dB
Decay time	D-RT	μs
Rise angle	A/RT	dB/μs
Decay Angle	A/(D-RT)	dB/μs
Rise Time/Decay Time	RT/(D-RT)	-
Energy/Amplitude	E/A	attoJ/dB
Counts to peak/Counts	CP/C	-
Amplitude/Frequency	A/AF	dB/kHz

Table 2. Thirty features extracted from the waveforms in time, frequency and wavelets domains

	Descriptor	Unit
Time domain	Amplitude	*V*
	Duration	*s*
	Energy	*V*2
	Zero-crossing rate	%
	Rise time	*s*
	Temporal centroid	*s*
	Temporal decrease	*V.s*−1
Frequency domain	Partial Power 1	%
	Partial Power 2	%
	Partial Power 3	%
	Partial Power 4	%
	Frequency centroid	*Hz*
	Peak frequency	*Hz*
	Spectral spread	*Hz*
	Spectral skewness	-
	Spectral kurtosis	-
	Spectral slope	*Hz*−1
	Roll-off frequency	*Hz*
	Spectral spread to peak	*Hz*
	Spectral skewness to peak	-
	Spectral kurtosis to peak	-
	Roll-on frequency	*Hz*
Wavelets domain	Wavelet Packet Energy 1 (LLL)	%
	Wavelet Packet Energy 2 (LLH)	%
	Wavelet Packet Energy 3 (LHL)	%
	Wavelet Packet Energy 4 (LHH)	%
	Wavelet Packet Energy 5 (HLL)	%
	Wavelet Packet Energy 6 (HLH)	%
	Wavelet Packet Energy 7 (HHL)	%
	Wavelet Packet Energy 8 (HHH)	%

The correlation matrix of the descriptors is calculated and subjected to complete link hierarchical clustering [35, 49, and 50]. The correlation matrix of the AE descriptors are obtained by evaluating the degree of linear correlation between each couple of descriptors, using the Pearson's coefficient *r*:

$$r(\bar{X},\bar{Y}) = \frac{1}{n}.\sum_{i=1}^{n} \bar{x}_i \bar{y}_i \quad r \in [-1;1] \tag{5}$$

Where $\bar{X}$ and $\bar{Y}$ are two standardized vectors and n is the total number of coordinates, *r* is equal to zero when $\bar{X}$ and $\bar{Y}$ are uncorrelated.

The resulting dendrogram allowed for the elimination of redundant descriptors (Figure 4). To keep significant information about each signal, a maximum correlation coefficient 0.75 [30, 31, 42] is chosen, which corresponds to a non-correlation coefficient equal to 0.25. For this level of non-correlation, the less correlated descriptors are kept. Then a Principal Component Analysis (PCA) is performed in order to define new uncorrelated descriptors obtained from the linear combination of the selected descriptors and to reduce the data set size. At the end, the data is expressed in a 4-dimensional orthogonal basis.

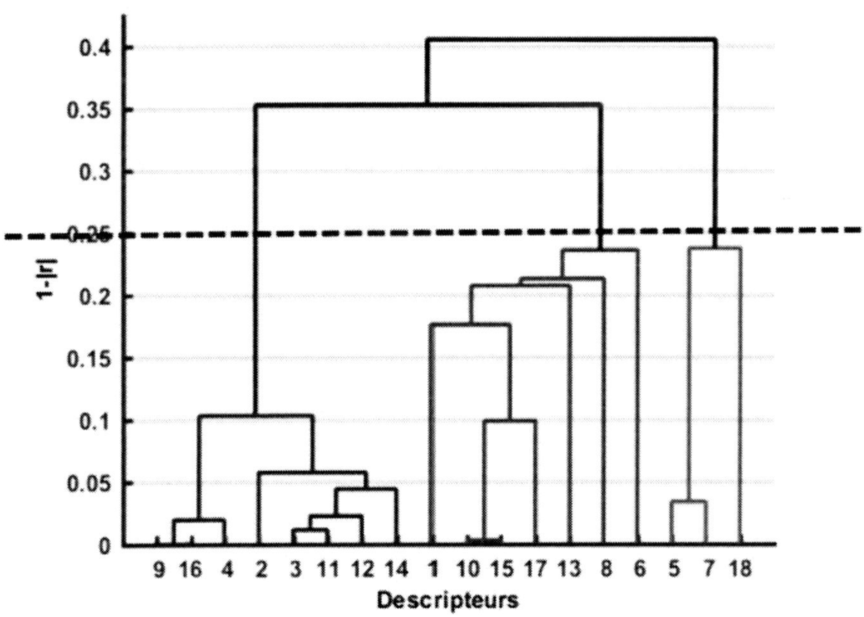

Figure 4. Dendrogram for the descriptors described in Table 2.

The first Principal Component (PC) is a linear combination of the uncorrelated descriptors, which accounts for most of the variability. Then, depending on data variability, additional components are introduced. Some AE events exhibit atypical characteristics. In the principal components space, they are very far from the majority of the data points. They are considered as outliers and eliminated in order to not influence the k-means clustering, which is very sensitive to such outliers.

In order to take into account that the principal components do not contain the same quantity of information, the distance between two points in this j D-space can be defined as:

$$d_w^2(X,Y) = \sum_{i=1}^{j} \lambda_i (X_i - Y_i)^2 \tag{6}$$

Where λ_i is the ith eigenvalue, X_i and Y_i are the ith coordinates of vectors X and Y.

2.2. Unsupervised Classification

Clustering refers to unsupervised classification. If no labeled data is available, an unsupervised methodology is required to perform an exploratory pattern analysis. "Unsupervised pattern recognition" is used to describe the complete methodology consisting of procedures for descriptors selection, cluster analysis and cluster validity. Partitional clustering methodologies produce a single partition of data for a prespecified number of clusters. AE signals with similar characteristics are segmented into clusters (which reveals the natural structure of data) based on similarity measures. Figure 1 summarizes the steps for the clustering. The k-means algorithm is one of the most widely used methods to determine cluster solutions for a particular user-defined number of clusters. The k-means method [51, 52] consists in minimizing the sum of squared distances from all vectors of a cluster to its center. This method assumes that the cluster number k is known and specified in advance. The k-means algorithm can then be described as follows:

1) Initialize the coordinates X_i of cluster center for each class C_i

2) Assign each descriptor vector X to a class Ci such as the weighted Euclidean distance between X and the cluster center X_i is the lowest among all the cluster centers
3) Compute the coordinates of the new cluster centers
4) If there is no change in the co-ordinates of the cluster centers, the algorithm has converged and the procedure is done; otherwise go to step 2.

One major problem in performing a cluster analysis with this method is to determine the optimal number of clusters for a given data set. To evaluate the quality of cluster partitions, several criteria can be used [53-55] such as the Davies-Bouldin validity index DB [56]. In general, it is assumed that the optimal number of clusters should be compacted within the cluster and well separated from other clusters. Therefore, this approach requires running the algorithm several times and the clustering partitions that optimize the validity index are selected as the best partition.

$$DB = \frac{1}{k}\sum_{i=1}^{k}\max_{i \neq j}\left\{\frac{d_i + d_j}{D_{ij}}\right\} \tag{7}$$

Where d_i and d_j are the average within-cluster distances of clusters, i and j respectively, and D_{ij} denotes the distance between the two clusters i and j. Therefore, several clustering steps are performed with k varying from 2 to 10. The best clustering result corresponds to minimum of the DB index (Figure 5a). Then, the Silhouette index [57] can be also used to evaluate the quality of clustering.

The silhouette S(i) value for each AE signal is a measurement of how similar this signal is to the other signals of its own cluster compared to the signals in other clusters. It is defined as:

$$S(i) = \frac{b(i) - a(i)}{\max\left[b(i), a(i)\right]} \tag{8}$$

with

$$b(i) = \min_{k} B(i,k) \tag{9}$$

where a(i) is the average distance between the ith AE signal and the other signals of its own cluster, and B(i, k) is the average distance between the ith AE signal and signals in another cluster k. This measure ranges from +1, indicating AE signals very distant from neighboring clusters, through 0, indicating AE signals not distinctly in one cluster or another, to -1, indicating AE signals probably assigned to the wrong cluster. The silhouette of a cluster is a plot of S(i) distribution for all AE signals in the cluster. Global measures of the silhouettes are given by averaging those of the entire data set.

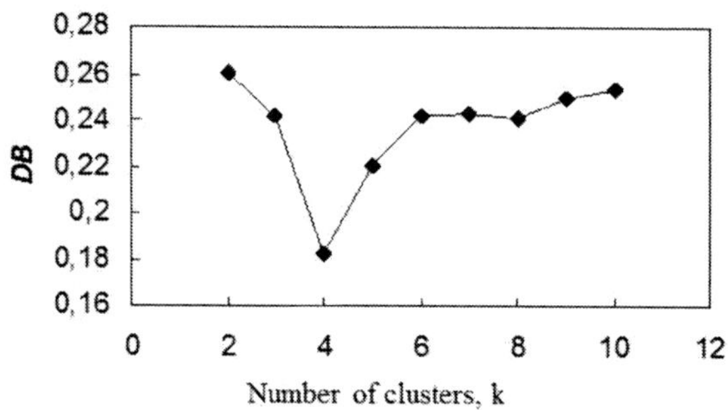

a)

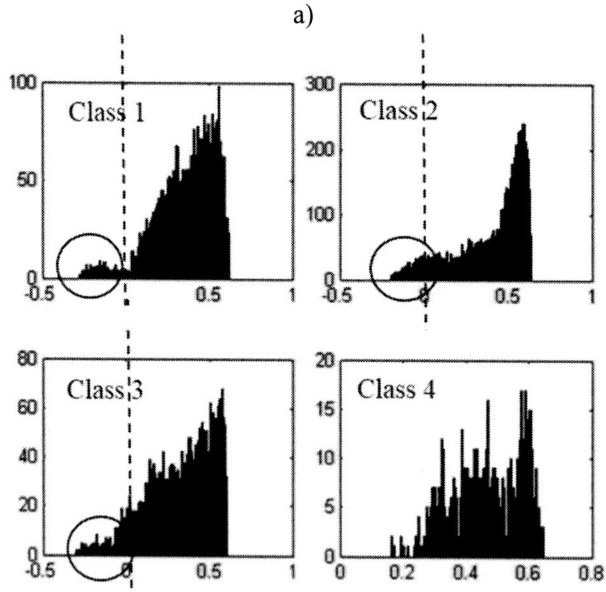

b)

Figure 5. a) Evolution of the DB values with the number of classes, the optimal solution is obtained with k=4 b) Histogram of the individual silhouettes for each signal recorded in the four classes showing signals with a negative silhouette.

The average silhouette width for the entire data set noted S(k) (m signals, k clusters) is defined as the mean value of all individual silhouettes, i.e., if m denotes the number of signals in the data set:

$$S(k) = \frac{\sum_{i=1}^{m} s(i)}{m} \tag{10}$$

S(k) characterizes more broadly the heterogeneity and isolation properties of the cluster k. This coefficient is used as a parameter to evaluate the best clustering. The global Silhouette can be used as an effective validity index. Moreover, the signals having a negative silhouette can be eliminated from the analysis because they are badly classified (Figure 5b).

Unfortunately, the k-means algorithm can reach either a local or a global convergence depending on the initialization. Sibil [43], in order to improve this algorithm, used a genetic algorithm with the k-means method. The algorithm starts typically with a set of solutions (randomly generated) called the population and creates successive new populations by genetic operations generations such as natural selection, crossover and mutation. This algorithm's advantage is to scan the entire solution space. In this context, this is not only one solution, which is initialized, but also a set of solutions. The processes of clustering, fitness computation, selection, crossover and mutation are not executed for a maximum number of iterations. The decision of the classification is done with the convergence criterion for evaluation. The algorithm converged when variation per generation is less than 10^{-7}.

The most challenging point of the unsupervised pattern recognition technique is to identify damage mechanisms corresponding to different clusters. This is because the classification process does not lead to a unique solution and it does not exist an indisputable criterion to determine which classification result is more appropriate and representative of the damage mechanisms. The link between clusters and damage mechanisms is specific to each material and must be developed for each study based on the knowledge of the material.

2.3. Supervised Classification

Classification is the process of identifying the source of an event by comparing its features to those of previous events whose sources are known.

For example, k-nearest neighbor (KNN) [58, 59] rule classifies an AE signal as follows: 1) find the k closest signals in the training set, 2) predict the class by majority vote, i.e., choose the class that is the most common among those k neighbors. The neighborhood is evaluated using the lowest Euclidean distance. Optimum number K is determined using a cross-validation procedure (method of the "leave one out"). Each signal in the training set is considered as an unknown. For each value of K, the method is:

1) Split the training set into two parts A and B, where B contains only one vector.
2) Apply the K-NN method to the single vector B and classify B by its K-nearest neighbors in A.
3) Determine if B is correctly classified.
4) Repeat steps 1 to 3 for every possible choice of B.
5) Calculate the percentage of correctly classified vectors (average classification rate).

The optimum number K of the nearest neighbors is the one that minimizes the classification error.

With this supervised approach, the final decision can be obtained in two different ways. The first one is simply the usual majority voting (MV) rule. An unknown signal is classified according to the dominant membership of the K-Nearest Neighbors. Nevertheless, it is preferable to provide a confidence level for the final decision. Another decision rule is introduced called security voting (SV) rule. In this special rule, one given AE signal is assigned to a specific class if for example more than 70% of the total number of the K-Nearest Neighbors voted for that class.

Other algorithms require a learning phase such as the Kohonen map (Kohonen 1984) or the Random Forest algorithm [61, 45]. Yang et al. [62] show the significant potential for AE signal classification by the self-organization neural network technique. Kohonen's self-organizing map is a network of simple processing units called neurons. The Kohonen's algorithm is expected to topologically organize the AE data on a two-dimensional array network in such a way that the more the signals are similar, the closer they are represented on the map. The network architecture described in several papers [28, 40, and 41] consists of a two-dimensional array of 32x32 units. The learning phase for Kohonen's self-organizing map can be summarized as follows:

1) Initialize all the weights to a random value between 0 and 1.
2) Randomly select an input vector **X**; present it to the all neurons of the network and evaluate the corresponding quadratic distance output d_j.
3) Select the neuron with the minimum output d_j as the winning neuron i.e., the nearest vector to the input vector.
4) Update the weight of the winning neuron such that the winning neuron is as similar as possible to the input vector.
5) The neighbors of that neuron, defined by the neighborhood function are also updated.

This neighborhood function defines how many neurons, in the neighborhood of the winning one, are updated for each learning input. This parameter is highly important, being the basis of the map's topology forming. A convenient initial neighboring "radius" is found to be of 13, decreasing with k. As the training process is going on, the size of the neighborhood is decreased and at the end of the training, only the winning unit is adjusted.

1) Repeat steps 2 to 5 until all the input vectors X have been used at least one time.

After the learning phase, the Kohonen's map could then be used to monitor the chronology of the damage process during the lifetime of structural components. For any new AE signal, its Euclidean distance to each neuron on the map is calculated in order to assign to it the class of the closest neuron.

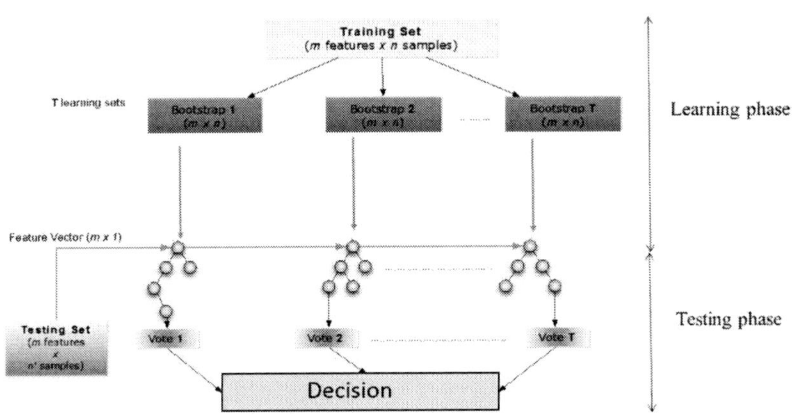

Figure 6. Schematic diagram of the Random Forests algorithm.

Random Forests (RF) are a set learning method which operates by constructing a multitude of decision trees during training, each capable of producing a vote when presented with a new set of features during testing. Breiman [62] originally develops the algorithm. The method merges two important ideas: the bagging and the random split selection of features in order to construct a collection of decision trees. Bagging (bootstrap aggregation) consists in building multiple training subsets by sampling with replacement (also known as bootstrapping) from the original training set. One tree is constructed per bootstrap (Figure 6). The second randomness phenomenon is introduced through the random split selection idea: at each node of the tree, only a random subset of all features is considered. During the testing phase, each AE signal is running down each tree of the Forest, leading to T votes.

3. Interest in the Fields of Materials: Examples

3.1. Martensitic Transformation

The aim of this work is to characterize plasticity-induced martensite formation of metastable austenitic stainless steel at room temperature [3, 63]. Acoustic Emission (AE) monitoring is performed on a 304L austenitic stainless steel during fatigue tests. This work aims at identifying the acoustic emission signals associated with the formation of the strain-induced martensite. Therefore, the sources of AE are: 1) cooperative dislocation motion 2) martensitic transformation 3) mechanical damage such as crack initiation and propagation. The AE signals have thus been classified into three clusters by using the k-means pattern recognition method. Cluster 1 signals have a low frequency with a long rise time. Cluster 2 signals have a shorter rise time with a higher frequency. The last cluster is constituted by very short signals. Cluster 3 corresponds to almost impulse signals. Figure 7 shows the projection of the classified signals onto the ratio amplitude/rise time versus average frequency plane. The three clusters are easily distinguishable in this plane, which means the algorithm succeeded in creating compact and well-separated clusters. The authors assume that the AE energy is proportional to the energy dissipated at the AE source, whatever the source mechanism.

The energy release of cluster 1 is very active at the beginning of the test during the first cycles. Nevertheless, at the end of the test, this cluster contains less than 30% of cumulated AE energy. As for the AE rate, this cluster is active throughout the test. Cluster 2 is increasingly activated as the test proceeds, reaching about 40% of the total number of signals and 60% of the total AE energy. Cluster 3 is not energetic (less than 10% of the total energy) and is activated by sudden bursts.

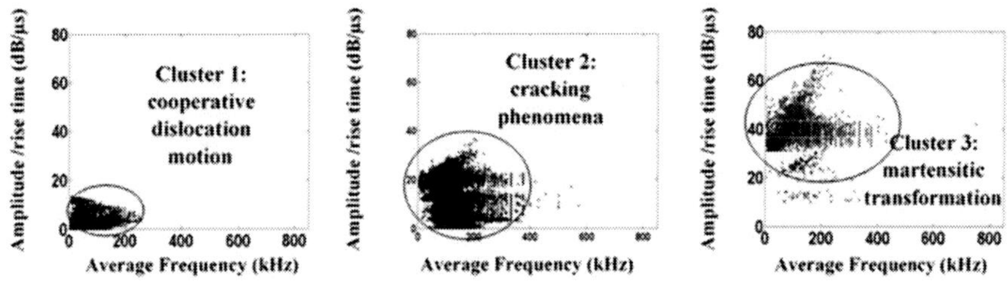

Figure 7. Clustering results in a 2D projection: amplitude/rise time ratio versus average frequency.

The analysis of AE clusters and identification of their mechanisms are very well explained in the article [3]. Cluster 1 is relied to collective dislocations movement, because the signals exhibit low energy, low frequency and long rise time. It is very active during the first cycle. This is consistent with the emission being generated by collective dislocation motion, very large during the first cycles due to overall plasticity. The decrease in AE activity of cluster 1 after the first cycle can then be linked with dislocation storage that reduces the mean free path of mobile dislocations.

Cluster 2 is the most energetic, containing short high-frequency bursts, releasing locally an important power. It is an energetic mechanism also activated during the unloading, during the compressive phase and very active at the end of tests. A rapid crack growth produces signals with short rise time and high energy. The crack lips can rub against each other in the compression phases, thus generating more AE. Indeed, cluster 2 is also activated during the unloading and the compressive phase. Moreover, it has strong activity at the end of the tests. All these arguments are coherent with the attribution of cluster 2 to the cracking phenomena.

The martensitic laths observed are very fine so that the expected released acoustic energy accompanying martensitic transformation (MT) is weak but intense, due to the displaced character of MT, MT should produce delta like signals. Indeed, the analysis of the AE cluster 3 shows that the emission is composed of very short signals. The energy involved in cluster 3 seems to be proportional to the rate of MT. The statistical distribution of energies for the signals of cluster 3 are found to follow a power law over 4 decades with no sign of cut-off. The exponent $\alpha = 1.75 \pm 0.15$ for the power law is consistent with previous experimental or theoretical works on critical dynamics [64]. Moreover, a lone superposition is found between the cumulated energy curve of cluster 3 of AE with an Eddy Current modulus curve (Figure 8). Therefore, Cluster 3 corresponds to martensite formation.

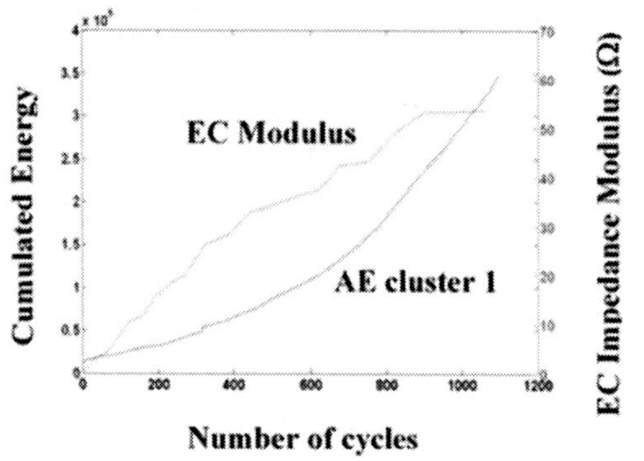

a)

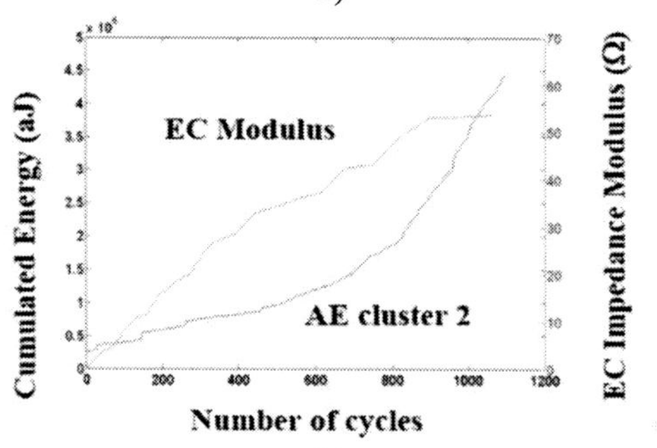

b)

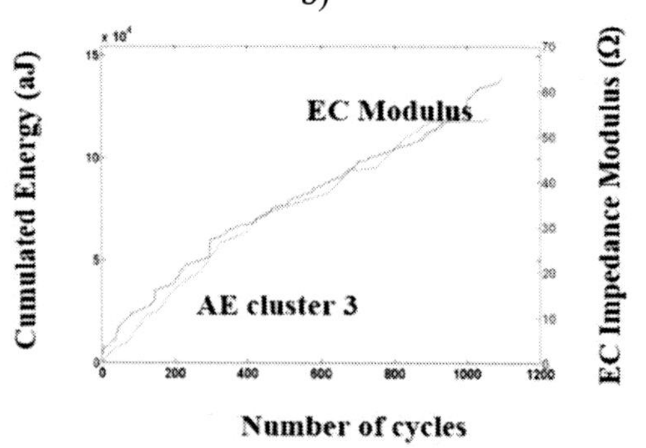

c)

Figure 8. Evolution of the Eddy Current (EC) modulus and the cumulated energy of each cluster: cluster 1 (a) cluster 2 (b) and cluster 3 (c) versus the number of cycles.

Taking into account the conclusions obtained from AE activity, each cluster has been correlated with a different AE mechanism as follows: (i) signals in Cluster 1 are assigned to dislocations; (ii) signals in Cluster 2 to mechanical damage and (iii) signals in Cluster 3 to martensitic transformation. Most remarkably, the statistical distribution of energies for the signals attributed to the martensitic transformation are found to follow a power law over four decades.

Altogether, the k-means classification proved to be a powerful tool to find order or at least universality behind the apparent disorder of concurrent AE mechanisms.

3.2. Refractories: Investigation of the Mechanisms of Cracking by Acoustic Emission

High zirconia refractories, mainly used in glass furnaces, present a very atypical structure. Their manufacturing process, by casting, leads to a zirconia skeleton (>90% in mass) made of interconnected dendrites surrounded by a so-called intergranular glassy phase (Sibil et al. 2009). During their cooling from high temperatures (>2000°C), these materials undergo two successive phase transformations: cubic to tetragonal and tetragonal to monoclinic at about 1000°C, this temperature depending on the presence and the quantity of a zirconia stabilizing agent. Acoustic emission monitoring has been performed during cooling from 1200°C. The refractory sample is held in a furnace by two zirconia waveguides located on each side of its length. Thanks to this configuration, a linear localization of the signals is performed to ignore noise (Sibil et al. 2011). The results show that there is no activity above the glassy phase transition temperature except for the tetragonal to monoclinic phase transition. This could be explained by the behavior of the glassy phase that accommodates the evolutions of the zirconia skeleton. On the contrary, the energy released by signals increases in an exponential way as the temperature decreases, indicating that cracking is more and more severe (Figure 9). Though it has been emphasized by acoustic emission monitoring that damage during cooling occurs mainly after the second transformation, there is still a lack of data concerning the mechanisms responsible for the observed cracking. Successive tests of segmentation are thus carried out for a varying number of classes from 2 to 10. The two criteria of validation, coefficient of Davies and Bouldin and the silhouettes indicate very clearly that the optimal solution contains two classes. The two classes present very unbalanced populations:

51,190 signals for the class n°1 (in blue) against 2,391 signals for the class n°2 (in green). The representation of the signals in the planes (time/temperature and energy/temperature) makes it possible to specify the manifestation of the damage (Figure 9). With cooling, damage is observed at around 890°C, temperature corresponding to the vitreous transition from the intergranular vitreous phase. Initially materialized only by the signals of the class n°1, it gathers, around 200°C, the signals of the two classes. This double contribution at low temperature marks, besides the strongest energies for the signals recorded in the class n°2, an increase of damage. The authors formulate the assumption according to which both reflect a mechanism of cracking but are characterized by the fact that this cracking takes place on different levels of activation. The assessment by XRD of the thermal expansion coefficients of zirconia at the lattice scale and the analysis of EBSD maps show that cracking is produced by the thermal expansion mismatch between groups of crystallographic variants. It can be noticed that cracks propagate into two modes: either they propagate along the tetragonal domains or they crossed a tetragonal domain almost perpendicular to the neighboring domains. Moreover, they are systematically initiated at the interface of two tetragonal domains. With regard to the propagation of cracks, the study shows that cracking within a tetragonal domain can be stopped or deviated by the presence of one of the other variants of the corresponding quadruplet. This seems to indicate that the crossing of another variant involves a large quantity of energy.

3.3. Composite Materials

This technique has been used to characterize damage evolution in different types of glass fibers-polyester samples, which are expected to produce preferential damage mechanisms [40]. Firstly, pure resin samples and 90° and 45° off-axis unidirectional composite samples are used. Tensile tests are performed in order to generate expected damage modes: matrix fracture in pure resin samples, mainly matrix fracture with a few decohesions in 90° off-axis samples and mainly decohesions with some matrix fracture in 45° off-axis ones. Two main types of signals are identified, denoted A and B, coming from the two damage mechanisms: matrix cracking (A) and fiber/matrix decohesions (B) (Figure 10a and Figure 10b). The initiation, the growth of matrix cracks and their coalescence appear to create AE events having low amplitude, slowly rising and decaying. On the contrary, the interface debonding leads to signals with higher amplitude, quickly rising and decaying. However, the difficulty in separating objectively two clusters within a large quantity of signals lead the authors to use neural network based pattern recognition techniques. So, the combination of acoustic emission amplitude distributions and neural networks, in the form of a Kohonen self-organizing map is successfully employed to discriminate signals originating from different damage types.

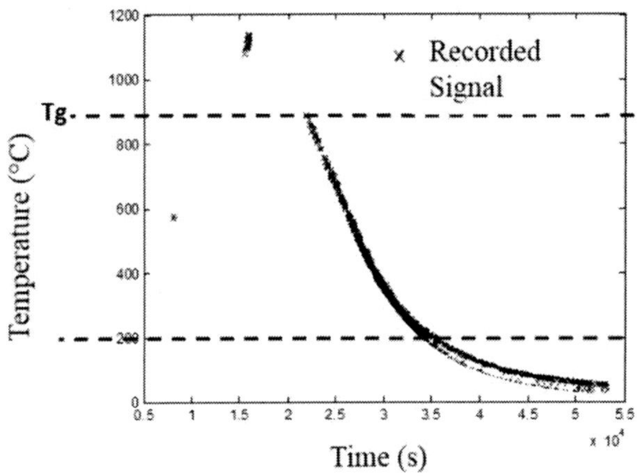

a)

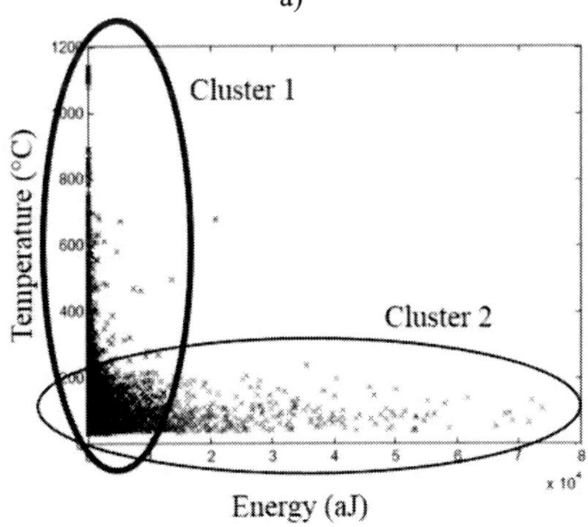

b)

Figure 9. Acoustic emission monitoring during heating and cooling: (a) Represented in the plane temperature versus time (each blue point corresponds to a recorded signal) and (b) Energy of the recorded signals in cluster 1 and 2 versus temperature.

Table 3. Results of the activation of the Kohonen's map for the identification of fibers breaks, matrix cracking and debonding between fiber and matrix during tensile tests

	C-type signal Fiber breaks	A-type signals Matrix cracking	B-type signal Decohesion
Tensile test on pure resin: Polyester Resin	0%	**76%**	24%
90 ° off-axis tensile test	5%	**60%**	35%
45 ° off-axis tensile test	8%	12%	**85%**
Bundle fiber: Glass Fibers	**70%**	5%	25%
Single fiber composite	**55%**	5%	35%

At a macroscopic level, the cumulative plot of the events of each class versus time clearly indicates damage initiation and progression of each failure mechanism. The interpretation of clustering results on a macroscopic level is performed by plotting the cumulative events of each class versus load. It indicates a typical evolution of the AE event, for the discriminated source mechanisms. For the 90° sample, it is observed that the two classes exhibit similar activity trends. The fraction of A-type signals for the 90° sample evolved more prominently than for the 45° sample, from the beginning to the end of deformation. On the contrary, a larger fraction of B-type signal is detected for the 45° than for which decohesion is the dominant damage mechanism. Let us also note that the number of B-type signals associated with decohesion greatly increases at the end of tests, causing the final failure of the material. Moreover, Kohonen's map is successfully used to establish the chronologies of the different damage modes at any stage of the material ageing [41] (Figure 10c). Specimens were immersed in deionized water above room temperature (50°C or 70°C) in order to accelerate the hydrolytic ageing. Compared to the case of unaged materials, the predominant type of damage is now interfacial debonding whatever the sample orientation. This work can lead to perspectives like real-time detection of damage in complex glass fibers reinforced polymers pipes structures, by analyzing the evolution of AE activity originating from the 'most dangerous' damage modes considered as the acoustic signature of the different damage modes. An extension is done; the previous training set has been enriched with AE signals issuing from fiber failure [28]. The work has been done on model composites made of a single E-glass filament embedded in a polyester resin. In a fragmentation test, the single fiber progressively breaks under increasing tensile strain into an increasing number of shorter and shorter fragments. Waveforms of signals referred to as C-Type signals are associated with fiber breaks. These signals of high amplitude are quickly rising and decaying. They are very similar to the B-type signals. The same approach is now used in order to test the classifier ability to sort signals resulting from three different types of damaging events; matrix cracking, interfacial decohesion and fiber failure. The classification results are summarized in Table 3.

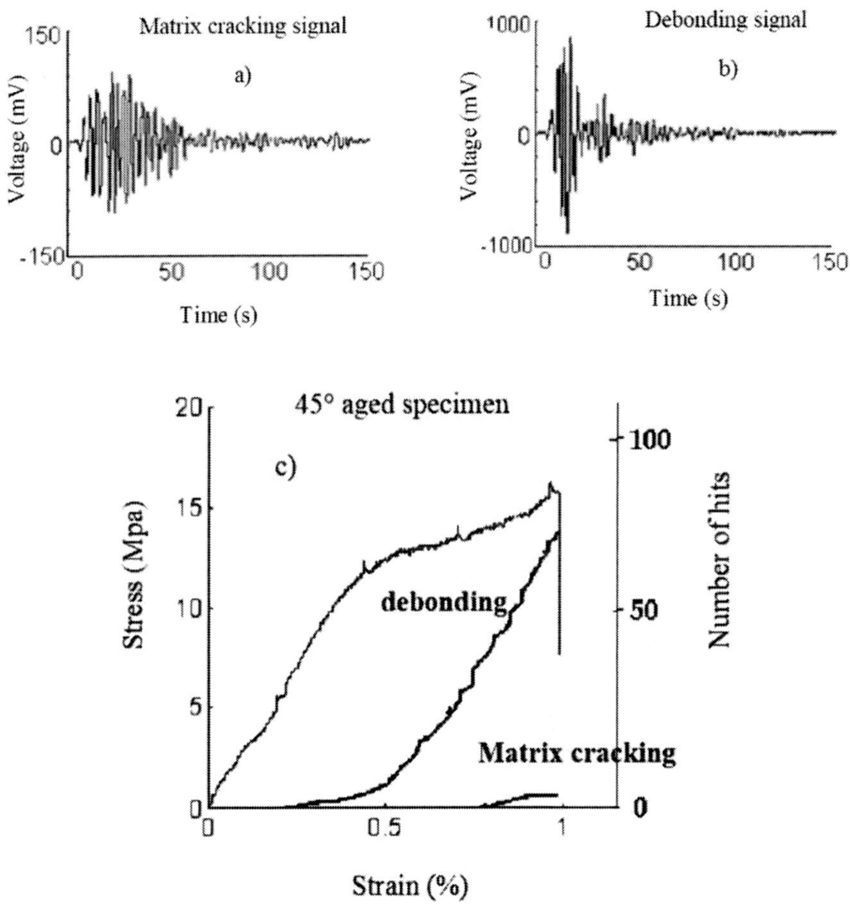

Figure 10. a) a representative A-type signal of matrix cracking and B-type signal of debonding collected during tensile test on 90° unaged sample b) Stress and cumulated number of signals for each mechanism versus strain, tensile test on 45° aged specimens.

Application of the classifier on AE data shows that it is difficult to distinguish AE signals due to fiber failure and AE signals due to decohesion. Again, most of the whole data collected during tensile tests on pure resin samples are associated to type A signals. Concerning the unidirectional composites, the signals are mainly distributed amongst type A and type B. However, 5% of the signals are assigned to type C though there is no mechanical explanation for the fiber failure in the case of 90° off-axis tension. Moreover, only 55% of the signals collected during tensile tests on single fiber samples are assigned to type C whereas 35% of them are assigned to type B.

About this behavior, the following hypothesis can be put forward:

i. The AE signatures of interfacial decohesion and of fiber, failure could be very similar or the descriptors used could be inadequate. In this case, it would probably be judicious to discern more pertinent descriptors;
ii. The fragmentation test of single fiber samples could give rise to quasi-simultaneous fiber failure and interfacial decohesion so that the resulting AE signal could be very confused.

Ceramic matrix composites (CMC) are interesting structural materials for high temperature applications, owing to their good mechanical properties at elevated temperatures [67-70]. C_f/SiC composites have been developed for aeronautic and aerospace applications [42]. Damage of fiber reinforced ceramic matrix composites involves several phenomena at various length scales including matrix cracking, fiber debonding, and fiber failures. A 3D multi-layered composite made of self-healing (Si-B-C) matrix reinforced ex-PAN (High Resistance) carbon fibers (40% volume fraction) is investigated. Test specimens with 4 mm thickness, 16 mm width, and 30 mm gauge length are tested. Static and cyclic fatigue tests are conducted in air at various temperatures (700°C, 1000°C and 1200°C) under uni-axial on axis tensile loading. The cyclic fatigue tests at 700°C and 1000°C are conducted under tensile/tensile sinusoidal loading with constant amplitude and 0.25 Hz frequency. Two methods for signal discrimination are implemented in order to analyze AE data and identify source mechanisms involved during fatigue at high temperatures on C_f/Si-B-C) composites, an unsupervised classification (PCA + k-means) and a supervised classification (K-nearest neighbors). First, the unsupervised classification gave reproducible clustering solutions with 4 (denoted A, B, C and D) or 5 classes (denoted A, A', B, C and D) in static fatigue (depending on testing conditions) and 4 (denoted A, B, C' and D) classes in cyclic fatigue. The analysis of AE activity, mechanical behavior and SEM observations led to the identification of six distinct types of AE signals. These signals are associated with one main damage mechanism: Class A: collective fiber breaks, Class A': individual fiber breaks, Class B: matrix cracking, Class C: fiber/matrix debonding, class C': yarn/yarn debonding, Class D: sliding at fiber/matrix interfaces and closure of matrix cracks after unloading. The training set for the supervised classification set is created by merging AE data collected during reference static fatigue tests (1200°C-150 MPa) and cyclic fatigue tests (700°C-0/130MPa). The library of signals is then used to identify the damage modes generated during fatigue tests performed at various temperatures. During fatigue tests at high temperatures, it is observed that in addition to classes A, B, C and D, a small amount of type A' and type C' signals is identified (Figure 11). This effect can be explained by the fact that at 700°C a few fiber breaks and inter tow debonding occurred in static fatigue. The results obtained using the supervised classification technique indicated that the fraction of type A' signals increased with temperature in static fatigue (2% to 5% at 700°C and 1000°C and about 15% at 1200°C) and the fraction of type C' signals increased under cyclic fatigue (1% to 6% in static fatigue and about 17% in cyclic fatigue). At 700°C, the amount of type A signals is significantly greater than the A' ones. Under static fatigue, the amount of type-C' signals is very low, whereas in cyclic fatigue it became predominant compared to the Class C signals. These results are therefore more satisfactory than those obtained using the unsupervised classification, since individual fiber failures, collective fiber failures, fiber debonding, bundles debonding can be distinguished at whatever temperature, load level and loading mode (static or cyclic fatigue). The use of a supervised classification allowed identifying AE signals with a better accuracy regardless of testing conditions (temperature, applied stress and loading mode). The six types of signals could be distinguished even when present in small quantities. It allows real-time identification of damage mechanisms regardless of testing conditions (temperature, applied load and loading mode).

3.4. Application to Localized Corrosion

Considering the crevice corrosion process, emitted gas (i.e., bubbles) coming from chemical reactions generate AE activity, which can be recorded by sensors located on the surface of the specimen. Since AE signals associated with crevice corrosion are characterized by low energy content, it is very difficult to separate those signals from the environmental noise [71, 72]. The chosen classification algorithm must be fast, reliable and not very sensitive to a mislabeled learning database (due to real-time and reliability industrial constraints). Moreover, it is preferable to provide a confidence level for the final decision. A whole approach combining the waveform preprocessing and Random Forest supervised classification has been implemented [45]. The final decision can be obtained with the majority voting (MV) rule and with the security voting (SV) rule. In these special rules, one given AE signal is assigned to a specific class if more than 70% of the total number of trees voted for that class. The whole approach combining the waveform preprocessing and the Random Forest supervised classification has been implemented into the software RF-CAM ("Random Forests Classification for Acoustic Emission Monitoring"). Experiments are conducted on 304L stainless steel. Samples are immersed in a corrosive solution (with different values of NaCl (2g/L or 35g/L), pH (6.7, 8.3 or 10.5) and temperature (25°C or 50°C). Some experiments are performed with the addition of H_2O_2 in order to accelerate corrosion. The Open Circuit Corrosion Potential (OCP) drop shows that crevice initiates as soon as H_2O_2 is added. In order to build various data sets, pH and temperature values, NaCl concentration and H_2O_2 addition are controlled to obtain crevice corrosion for some experiments and no corrosion for the others. In absence of corrosion, OCP does not decrease. However, even if there is no corrosion, some AE activity is observed all along the experiment. These signals are attributed to noise (i.e., dilatation phenomena and bubbles evolution within the liquid due to the testing temperature of the test (50°C)). A first class (denoted as NC) is built from 1200 signals. Regarding the experiments involving crevice corrosion, AE activity starts prior to the addition of H_2O_2, and is mainly assigned to noise. After H_2O_2 addition, gathered signals are mainly assigned to corrosion and constitute a second class (denoted as CC). It has to be noticed that this CC class is not pure and contains noise signals. The remaining experiments are used to construct the different testing sets. One of them presents no corrosion (namely NCa); the other shows crevice corrosion (namely CCa). The percentage of signals for the Majority Voting (MV) rule is given regarding the original total number of signals of the test set. The percentage of signals for the Security Voting (SV) rule is given regarding the number of remaining signals, after the security threshold of 70% is applied.

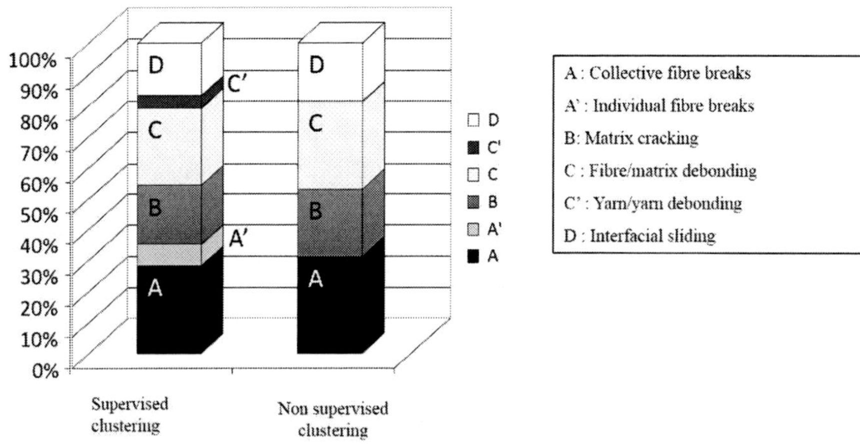

Figure 11. Comparison of signals proportion obtained for each class with the supervised classification and with the unsupervised clustering for the test carried out in static fatigue at 700°C.

For each test (Table 4), the proper majority class has been recognized and results show that using SV leads to a reinforcement of the usual MV decision, when it comes to assigning signals to a specific class. Moreover, for each test case, most of the signals corresponding to the minority class are discarded, thus strengthening the trend of the majority class. Future prospects to be considered are the enlargement of the learning library in order to identify other corrosion mechanisms such as pitting corrosion. Finally, this methodology, which is developed and validated at the laboratory scale, can be applied at the industrial scale, providing the construction of a consistent new learning library.

Table 4. Random Forest classification results for NC and CC experiments

Testing set	NC class Number of votes (%)	CC class Number of votes (%)	Decision rule
NCa	95.6% 98.3%	**4.4%** **1.7%**	**MV** **SV**
CCa	**34.9%** **29.3%**	65.1% 70.7%	**MV** **SV**

4. LIMITATIONS: TOWARDS A QUANTITATIVE APPROACH

The link between clusters and damage mechanisms is specific to each material and must be developed for each study based on knowledge of the material. This approach is based on empirical correlations between the characteristics of the sources and the measured AE signals or features that have been extracted from the signals. These correlations are established after repeated tests on identical measurement conditions.

Through the identification of damage mechanisms by AE using the unsupervised method, it is sometimes difficult to validate the classification and labelling of AE data. Indeed, most of the time, different source mechanisms are activated simultaneously and in unknown proportions. After AE data clustering, the maximum number of clusters may be equal to the number of mechanisms if all acoustic signatures are distinguishable. But some mechanisms may produce similar AE signals that would not be separated by the clustering algorithm. In this case, the number of clusters would be lower than the number of mechanisms.

In most work on AE, the assignment of a signal to a damage mechanism is very difficult to validate, mostly impossible without a modelling work [73, 74]. Moreover, the wave originating from the source is altered during propagation. Hence, the main limitation of the AE techniques is that changes due to the acquisition chain are not well known. First, it is modified by the propagation medium itself [75-78] (phase displacement, reflections, dispersion) and then, by the recording system [79, 80]. All these transformations certainly make the interpretation of the signals very difficult. On the first part of this section, we are going to investigate, with ground truth data, the influence of the description of AE signals on the clustering. The second part is dedicated to the investigation of the influence of specimen size and AE signals acquisition.

4.1. On Ground Truth Data

The performance of the classifiers is evaluated using data sets for which the partitions are known. Data sets are generated digitally by creating and combining multiple clusters of signals [43, 45]. The signals from each cluster are described by a set of descriptors whose values are chosen close to those obtained from experimental data (monitoring on steel, ceramics, glass fibers, composites ...) and several clusters are created. The number of signals in each data set is 2000. All signals in a data set are expressed in an 18-dimensional space. The data sets contain separate and overlapping clusters.

4.1.1. Relevant Descriptors

The influence of the choice of descriptors is examined with a known data set, close to data resulting from mechanical tests. Indeed, after AE data acquisition, every AE signal is represented by 18 descriptors [31]. An important question is: is it necessary or useful to keep the 18 descriptors, or is it better to select relevant descriptors? In this study, the user according to the apparent structure of data defines the relevant parameters. So, this data set is initially described by 18 descriptors (and expressed in the 5-principal components' base) then by only 3 relevant descriptors (amplitude, ln(number of counts to peak), ln(energy)) and expressed in the 3-principal components' base.

When the 18 descriptors are considered, the best segmentation obtained is always a 3-cluster solution. Only Cluster 1 is correctly identified with 1985 signals. The average Silhouette for this cluster is greater than 0.6, whereas for the other two clusters, the average Silhouette is lower than 0.5. The clustering does not reveal the structure of the data. Considering only three relevant descriptors (amplitude, ln(number of counts to peak), ln(energy)), the 4-cluster solution corresponding to the data structure is obtained. With the k-means algorithm, the correct solution is obtained 8 times out of 10, whereas the genetic strategy obtained the correct solution 10 times out of 10. The four clusters had an average silhouette greater than 0.6 and there are only 120 misclassified signals.

This result shows that the choice of the input descriptors is essential. In unsupervised classification, it is impossible for the user to visually determine the relevant parameters. Thus, it will be useful to test different sets of descriptors and to use several criteria such as DB and S(k) criteria to select the best score [21].

4.1.2. Minor Mechanism

A major challenge is to distinguish a cluster if a source mechanism becomes minority. "Mechanical test inspired dataset," containing the same clusters as the previous set but with different sizes, is created. Cluster 1 still contained 2000 signals, Cluster 2 only 500 signals, Cluster 3, 1000 signals and Cluster 4 only 100 signals. In a mechanical test, damage mechanisms are activated very rarely in equal proportions. This data set will allow us to see if disproportion has an influence on the segmentation of data. This data set is described by only three relevant descriptors (amplitude, ln(number of counts to peak), ln (energy)). For the segmentation, data is expressed in the 3-principal components' base. With the k-means algorithm, the data structure is not revealed, indeed the two solutions obtained with k = 4 are incorrect; the 4 clusters which are created are not the good ones. It is interesting to note that the value of DB is 0.3 but the S(k) criterion never leads to a 4-cluster solution. Indeed, for these partitions, the value of S(k) is only 0.5. The majority cluster is split into two clusters and it appeared that the algorithm tends to create clusters with equivalent sizes. With the genetic strategy (9 times out of 10), both DB and S(k) criteria give a 4-cluster solution corresponding to the effective structure of data. The value of DB is 0.23 and S(k) is 0.65. The genetic strategy is less sensitive to different cluster sizes unlike the k-means algorithm. Cluster 1 contained 1969 instead of 2000, cluster 2, 551 signals instead of 500, cluster 3, 979 signals instead of 1000 signals and cluster 4, 151 signals instead of 100 signals. The error rate is only 3%. This result shows the interest of using several classifiers. Indeed, validity indexes may provide incorrect evaluation result in quantifying the separation distance between the centroids [81].

4.1.3. Outliers

Noise and outliers are unavoidable in AE monitoring. Sibil [43] shows the influence of the outliers on the results of clustering. "Outliers-cluster dataset," contained 4 clusters including a cluster with extreme values for its descriptors (Figure 12).

The relevant descriptors are the number of counts, signal duration and energy. The data set had 2000 signals per cluster. This data set will help answer this question: what happens if a source mechanism is very different from the others as outliers? For data set, which contained a cluster with extreme values, according to the DB index, both algorithms led to an optimal classification with two clusters (DB=0.17 and S(k)=0.67). The segmentation consisted of a cluster of 2000 signals (the average silhouette width of cluster 4: 0.96) and a cluster of 5937 signals (average silhouette for this cluster: 0.56). The two procedures isolated the extreme cluster and grouped the others together (Figure 12). The cluster validity indexes are a tool for assisting the user with the clustering process but they cannot be expected to solve all the problems related to the unsupervised classification. It is necessary to clean the data before clustering in order to isolate these outliers.

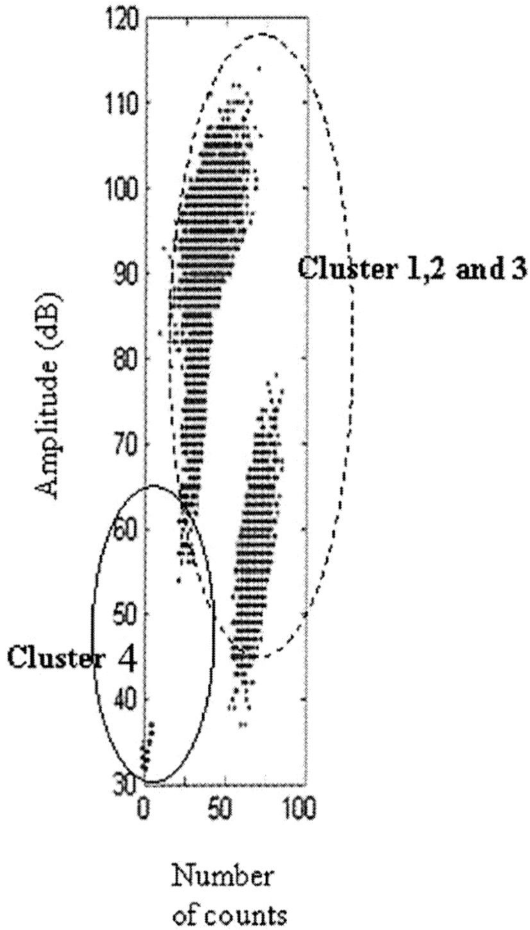

Figure 12. Visualization of data set with outliers, projection in a two-dimensional space (amplitude. number of counts) after partition.

Moreover [82] generates artificial clusters with several degrees of separation denoted J. For J equal to 0.01, the clusters are very close and for J equal to 0.342, the clusters are well separated. In all cases, the genetic-algorithm is able to find the partition with a suitable descriptor combination. Figure 13 shows the evolution of the cluster validity indexes for several problems (three clusters, four clusters and five clusters). We can notice that the values of the indexes evolve with the degree of separation between clusters.

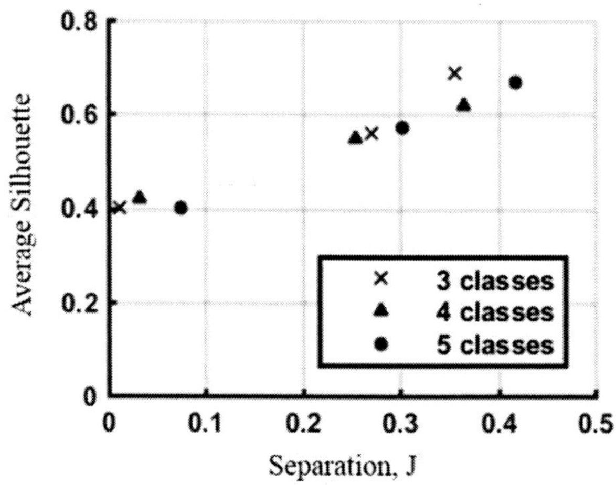

a)

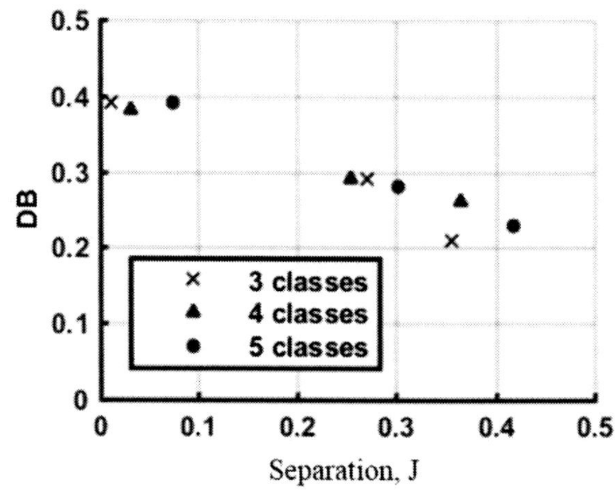

b)

Figure 13. Evolution of the indexes a) Davies and Bouldin b) Silhouettes with the degree of separation J between clusters for a three-class problem, four-class and five-class problem.

4.1.4. Influence of Mislabeled Data in the Training Set: Learning from Data with Uncertain Labels

The main assumption for a supervised classification is that labels of learning signals are true. Nevertheless, certainly, in the data set, some learning data have wrong labels. In this section, the results on artificial data sets illustrating the problem of supervised classification under uncertainty are presented. Those data represent four clearly identified classes (2000 signals per class). A training set is built, comprised of 70% of those data (5600 signals taken at random), the remaining 30% (2400 signals) constitute the Testing Set. This approach alters the original training set in order to simulate uncertainty on labelled data for supervised classification [45]. We can consider the general problem of label switching between the classes. For example, the class 1 becomes C1alter, composed of 79% of waveforms from class 1, 2% from Class 2, 15% from Class 3 and 4% from Class 4. In this case, the trust factor is equal to 80% or the contamination rate is equal to 20%. Figure 14 shows the evolution of the correct classification rate with the contamination rate. The recognition rate decreases with the contamination rate, 88% for a contamination rate equal to 30 and 99% for a trust factor equal to 100 for the k-NN algorithm. The Random Forest classification is more robust than K-NN for high contamination rates. The correct classification rate is equal to 96% for a contamination rate of 30%. Some authors propose a multi-class supervised classification method, called robust mixture discriminant analysis, for performing classification in the presence of label noise [83].

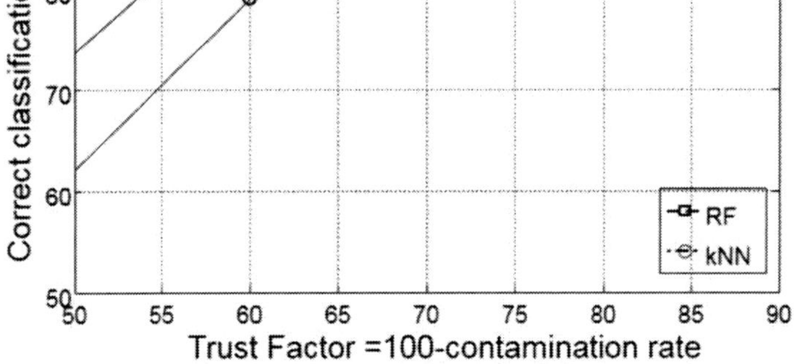

Figure 14. Performance of two supervised classifiers (Random Forest and K nearest neighbors) for increasing contamination rates on datasets.

4.2. Influence of the Specimen Size

In the Acousto-Ultrasonics technique (AU), the elastic waves are generated and recorded at defined locations by AE sensors. Thanks to the generation of controlled sources, the measures by acousto-ultrasounds allow us to investigate the effects of the propagation on the descriptors. Maillet 2012c and Maillet et al. 2014a investigate the influence of source-sensor and of dispersion and attenuation effects on ceramic matrix composite. The descriptors of the signals of AE are considered relevant if they are an image of the source independent from conditions of acquisition and from the degree of damage of the material. In this study, several types of signals are considered:

- Signals with variable rise time equal to 10, 20, 30 or 40 μs, maximal amplitude and frequency are controlled;
- Signals with constant amplitude and variable frequency between 150 and 950 kHz.

The rise time of the recorded signal is analyzed according to the rise time of the source for the various configurations. At low distance source - sensor (10 mm), there is a linear relation between the rise time of the recorded signal and the rise time of the source. On the other hand, for distances approximately 100 mm, the dispersive character of the material does not allow any source identification from information collected at the beginning of the signal.

Table 5. Frequency barycenter according to the frequency of the signal source and to the distance source- AE sensor for the specimen with thickness 1.5 mm (values in kHz)

	Distance source-sensor		
Frequency of the source	10 mm	100 mm	170 mm
150	156	152	153
300	298	301	302
450	452	438	437
600	597	594	591

Consequently, the rise time is not always representative of the source because it is very much affected by the propagation. Other descriptors (counts to peak, initial frequency, and rise angle) follow the same trend. The observation of the frequency contents of the recorded signals shows that in spite of the distance of propagation, it is possible to identify clearly the frequency of the source, independently of the thickness of the material and the rise of the source. Descriptors describing the frequency contents of the signals are a good image of the source (Table 5).

In the following study [82], a library is defined on a sample with a thickness equal to 3.7 mm and this library is then applied for data recorded on a sample a thickness of 1.5 mm. The only difference is the thickness of the specimen, 1.5 mm against 3.7 mm for reference. After the supervised classification with the K-NN algorithm, the number of mislabeled signals is very high (11.4% of the total number). It is due in particular to a high number of signals with a negative Silhouette value. Moreover, the results show that the identification of damage mechanisms on the thinnest sample from the library defined on the sample with 3.7 mm is not valid because the kinetics of appearance of the classes are not in accordance with expected mechanisms. It means that the acoustic signatures of the mechanisms of damage described by the library do not correspond to the acoustic signature of the same mechanisms within the set of data recorded on the thinnest specimen.

These results show the effect of the propagation and the sample thickness on the acoustic signature of damage mechanisms. Now, the effect of acquisition can be investigated.

4.3. Influence of the Acquisition

4.3.1. Coupling between Sensors and Material Surface

It is generally accepted that the energy of an AE signal represents a part of the energy released at the source. The recorded AE signal energy is affected by the distance of the wave propagation, energy attenuation due to damage, coupling between sensor and material, and by sensor frequency responses. The effects of attenuation due to propagation distance may be eliminated by combining AE signals energies recorded by sensors. But, the influence of coupling may lead to a different amount of energy received by each sensor even for a source located at an equal distance. This point may be investigated by comparing the amount of energy received at each sensor from any source located at mid-distance during tensile tests on a ceramic matrix composite. For comparison, the amounts of energy associated with sources located within a ± 5 mm distance from the middle of the gauge length are studied. In order to exclude the effects of damage, only the beginning of initial loading (strain lower than 0.1%) is considered. From the location of AE sources, it is checked that damage is limited and diffused for a strain lower than 0.1%. Therefore, the effects of damage are excluded as well. A comparison is then performed using the distribution functions of amounts of energy received by each sensor. It can be stated that both sensors being at the same distance from the domain, they should receive statistically the same proportion of signals energy. This means that both distribution functions should be superimposed when considering sources located in the domain. Figure 15 shows the distribution functions of energy for AE sources in the standardization domain (± 5 mm interval around the middle of the gauge length for a strain lower than 0.1%). The propagation distance to both sensors is quite identical. Effects of associated attenuation could thus be neglected. For some tests, the distribution functions are well superimposed (Figure 15a), whereas for others, a discrepancy is observed (Figure 15b), which is attributed to a difference in coupling between sensors and material surface. The authors [84] propose in-situ calibration in order to eliminate differences between AE sensors in terms of coupling to the material surface on the evaluation of the recorded energy. It is the same problem with the others descriptors but it is more difficult to correct them.

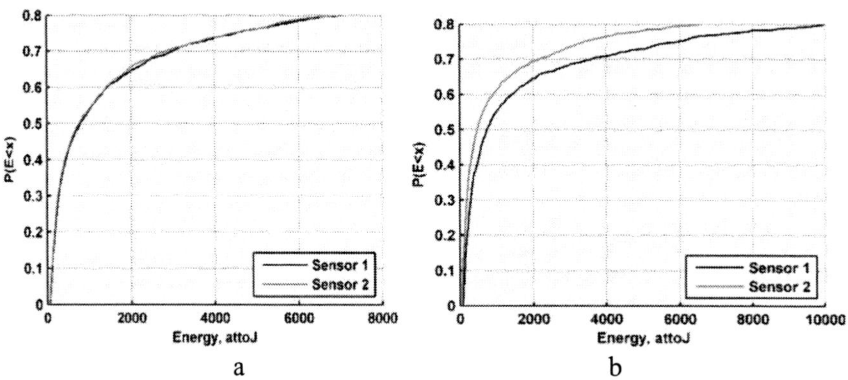

Figure 15. Energy cumulative distribution functions for two fatigue tests at intermediate temperature.

4.3.2. Position of the Sensors

The same mechanical tests are conducted on ceramic matrix composites with four sensors [85]. First, in configuration N°1, two resonant (micro 80, 250 kHz) sensors separated by 100 mm are put directly on the specimen on the gauge length, using medium viscosity vacuum grease as a coupling agent. In configuration N°2, two piezoelectric sensors (micro 80) are positioned inside the grips separated by 180 mm. The comparison of the recorded signals shows significant differences. The signals, which had a very short rise time in configuration N°1, are thus modified during their propagation. In configuration N°2 their rise time increased. The amplitude seems to decrease for all the signals, and the distribution is slightly moved towards the low values. (Figure 16).

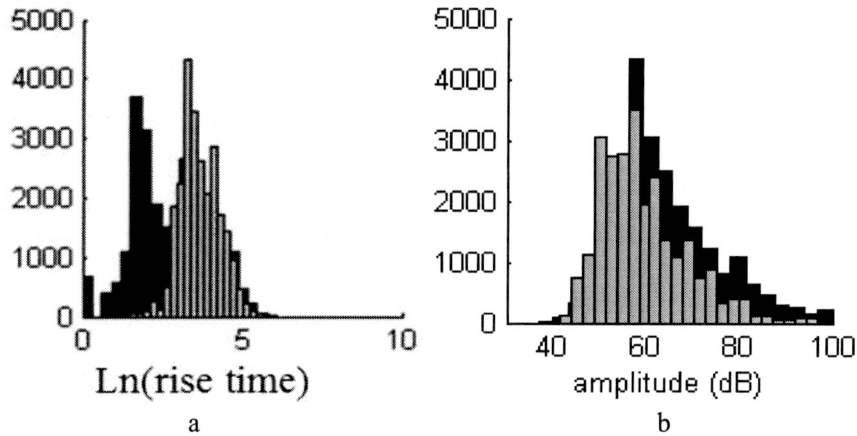

Figure 16. Comparison between the distributions of various descriptors of EA, for the data recorded in configuration n°1 (in black), and in configuration n°2 (in grey) on ceramic matrix composite for the same mechanical test.

For the data recorded in configuration N°1, the optimal number of clusters is equal to four denoted as A, B, C and D. (Table 6).

Table 6. Average Characteristics of the four classes of AE signals recorded on ceramic matrix composites (configuration n°1: sensors on the gauge length)

Class	A	B	C	D
Rise time (μs)	7	43	7	47
Number of counts	1011	1104	172	191
Duration (μs)	3194	3461	781	845
Amplitude (dB)	78	72	60	56
Av. frequency (kHz)	238	236	200	195
Counts to pic	4	16	4	16
Initial freq. (kHz)	473	328	458	309
Rise time/duration	0,001	0,012	0,009	0,056
Energy (attoJ)	53866	19371	754	480

Table 7. Average characteristics of the two classes of AE signals recorded on ceramic matrix composites (configuration n°2: sensors inside the grips)

Class	Energy (attoJ)	Amplitude (dB)	Rise time (µs)	Duration (µs)	Rise time/Duration	Average frequency (kHz)
K	5623	64	27	1835	0,01	237
L	556	54	55	1081	0,05	198

For the data recorded on the same specimen in configuration N°2, the optimal classification is obtained with only two classes denoted as K and L. The average characteristics of these two classes are given in Table 7. The class K has an average energy, an amplitude, a frequency and a duration bigger than class L, while its average time of rise is smaller. This result confirms the fact that we are less sensitive to the nature of mechanisms sources in configuration N°2 where we can distinguish only two shapes of signals instead of four in configuration N°1.

4.4. Contribution of the Modelling Work

Modal acoustic emission [86] is based on the concept of identifying the various waves excited within the sample due to the local release of energy. Gorman proposed this concept [76]. In this context, the finite element method is an ideally suited approach. Concerning the FE modelling, a 3D geometry has been built for several specimens denoted as EPL3, EPL16 and EPL80 [87]. The EPL16 specimen has the exact shape of the dumbbell specimen typically used in the laboratory for composite characterization. EPL3 is the size of a minicomposite specimen (only one bundle of fiber) and EPL 80 is the size of an industrial piece. In the three geometries (EPL3, EPL16 and EPL80), the same displacement history is introduced on the surface in order to rather equally simulate all frequencies up to 1.2 MHz. This simulation allows the author to identify the different modes of propagation by presenting the 2D Fast Fourier Transform (FFT2D). Those simulated modes are compared to theoretical modes calculated for plates and beams in the case of EPL80 and EPL3 specimens respectively. The results are presented as 2D Fast Fourier Transform shown in Figure 17. This way to present the results allows identifying the excited modes so that the simulated results may be compared to theoretical calculations. Here the results show that the propagation medium geometry affects the propagated signals. Knowing that, in these geometries, the signal propagates only following specific modes, it may be deduced that for the large specimen (EPL80) the plate modes are stimulated, while for the thin one (EPL3) the beam modes are stimulated. In the intermediate specimen (EPL16), for which the FFT2D is more complex, the identification of modes is practically impossible, as this geometry cannot be assigned either to a plate or to a beam. This study shows us that AE tests done on a classical specimen e.g., EPL16 in the laboratory cannot replace an industrial case from a geometry perspective. This result presents a good agreement with the results obtained by Gorman [88], which examined the transferability of the results from small laboratory specimens to larger tests. He concluded that the transfer is possible but a sound understanding of wave propagation and sensor response is required.

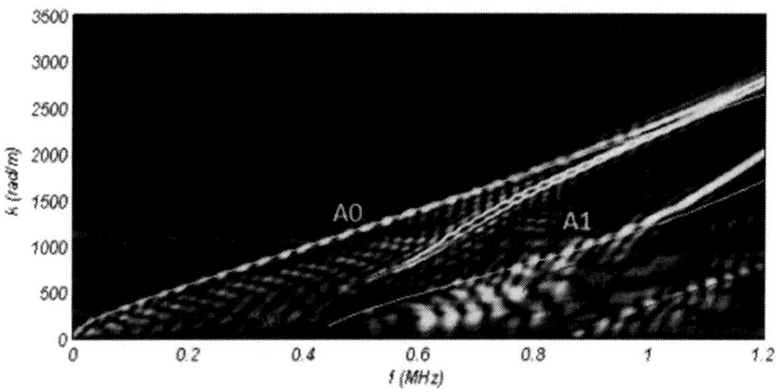

a)

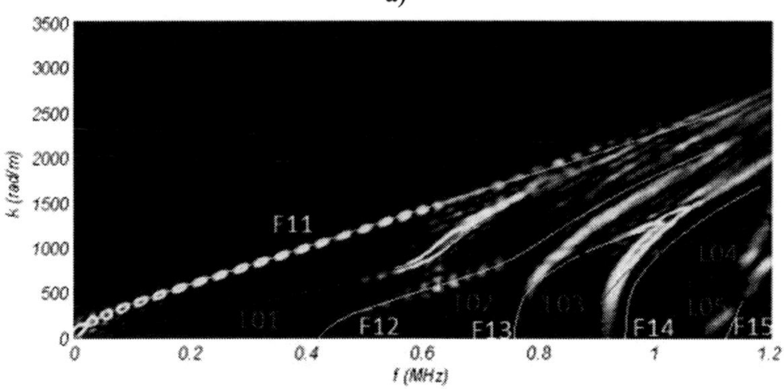

b)

Figure 17. (Continued)

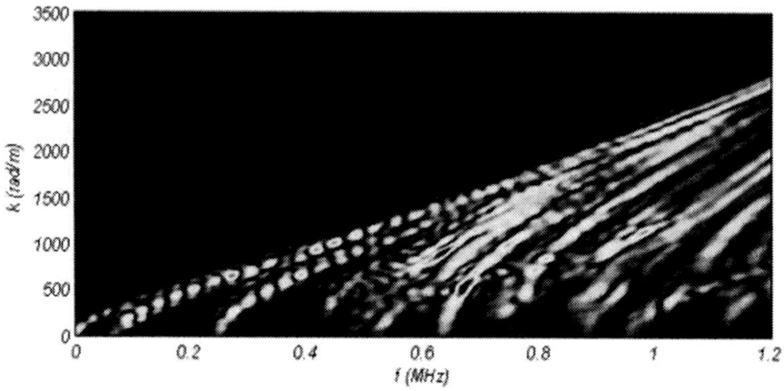

c)

Figure 17. FFT2D for the three geometries excited by the same source simulated on the surface, a) the large specimen (width = 80mm), b) the fine specimen (width=3mm) and c) the medium specimen (width=16mm) (Le Gall 2016).

Conclusion

This work can lead to outlooks like real-time detection of damage in structures, by analyzing the evolution of AE activity originating from the 'most dangerous' damage modes. In this industrial application on composite materials for example, where the tightness is required, matrix cracks and fiber-matrix decohesions are the main unexpected damages. As they allow the diffusion of water, they are considered as the most dangerous damages. The results obtained in terms of damage kinetics could now provide input data for mechanical models.

Each recorded signal depends on the source mechanism features, but also on material properties and sensor characteristics. In fact, AE signals are subjected to many effects caused by variations in the material along the raypath (heterogeneities, anisotropy) and by the recording system (coupling, sensor characteristic …). Therefore, the AE signal is the convolution of the source function, the material's Green function, the transfer function of the sensor and coupling and transfer function of the recording system. Specimen geometry influences elastic wave propagation and can make interpretation of AE signals difficult.

Accordingly, there is no universal signature for a given damage source. However, under defined set-up conditions, similarities exist between AE signals originating from similar damage. For an unsupervised clustering, there is a need to test the effectiveness and robustness of the validity indices on an arbitrary data structure in order to improve the clustering.

Nevertheless, through the identification of the damage mechanism by AE, it is difficult to validate the classification and labeling of AE data. In this kind of analysis, the assignment of a signal to a damage mechanism is very difficult, almost impossible to validate without a quantitative AE and modelling work.

REFERENCES

[1] Beattie, AG. 1983. "Acoustic emission, principles and instrumentation." *Journal of Acoustic Emission* 2:95-128.

[2] Pappas, Y.Z., Kontsos, A., Loutas, T.H., and Kostosopoulos, V. 1998. "Failure mechanisms analysis of 2D carbon/carbon using acoustic emission monitoring." *NDT & E International* 31:571-580.

[3] Shaira, M., Godin, N., Guy, P., Vanel, L., and Courbon, J. 2008. "Evaluation of the strain-induced martensitic transformation by acoustic emission monitoring in 304L austenitic stainless steel: Identification of the AE signature of the martensitic transformation and power-law statistics." *Materials Science and Engineering A* N°1-2, 492:392–399.

[4] Deschanel, S., Vanel, L., Vigier, G., Godin, N., and Ciliberto, S. 2006. "Statistical properties of microcracking in polyurethane foams under tensile test, influence of temperature and density." *International Journal of Fracture* 140(1-4):87-98.

[5] Deschanel, S., Vanel, L., Vigier, G., Godin, N., and Ciliberto, S. 2009. "Experimental study of crackling noise: conditions on power law scaling correlated to fracture precursors." *Journal of Statistical Mechanics* P01018.

[6] R'Mili, M., Moevus, M., and Godin, N. 2008. "Statistical fracture of E-glass fibers using a bundle tensile test and acoustic emission monitoring." *Comp Sci Tech.* 68:1800-1808.

[7] Maire, E., Carmona, V., Courbon, J., and Ludwig, W. 2007. "Fast X-ray tomography and acoustic emission study of damage in metals during continuous tensile tests." *Acta Materialia* Volume 55(20):6806-6815.

[8] Shiwa, M., Chen, O.Y., Kishi, T., Carpenter, S., Mitsuno, S., Ichikawa, H., Lee, Y.T., Kim, S.T., and Lee, T.S. 1995. "Fracture mechanisms in unnotched and notched SiC/SiC composites studied by acoustic emission analysis." *Mater Trans* 36(4):511-517.

[9] Siron, O., Chollon, G., Tsuda, H., Yamauchi, H., Maeda, K., and Kosaka, K. 2000. "Microstructural and mechanical properties of filled-added coal-tar pitch-based C/C composites: the damage and fracture process in correlation with AE waveform parameters." *Carbon* 38:1369-1389.

[10] Kaya, F. 2007. "Damage assessment of oxide fiber reinforced oxide ceramic matrix composites using acoustic emission." *Ceram Int* 33(2):279-284.

[11] Kim, S.-T., and Lee, Y.-T. 1997. "Characteristics of damage and fracture process of carbon fiber reinforced plastic under loading-unloading test by using AE method." *Materials Science and Engineering* A234-236:322-326.

[12] Kotsikos, G., Evans, J. T., Gibson, A.G., and Hale, J. 1999. "Use of acoustic emission to characterize corrosion fatigue damage accumulation in glass fiber reinforced polyester laminates." *Polymer Composites* 20(5):689-696.

[13] Kotsikos, G., Evans, J.T., Gibson, A.G., and Hale, J. 2000. "Environmentally enhanced fatigue damage in glass fiber reinforced composites characterised by acoustic emission." *Composites: Part A* 31:969-977.

[14] Ceysson, O., Salvia, M., and Vincent, L. 1996. "Damage mechanisms characterisation of carbon fiber/epoxy composite laminates by both electrical resistance measurements and acoustic emission analysis." *Scripta Materialia* 34(8):1273-1280.

[15] Barré, S., Benzeggagh, M. 1994. "On the use of acoustic emission to investigate damage mechanisms in glass-fiber-reinforced polypropylene." *Composite Science and Technology* 52(3):369-376.

[16] Margueres, Ph., Meraghni, F., and Benzeggagh, M.L. 2000. "Comparison of stiffness measurements and damage investigation techniques for a fatigued and post-fatigued GFRP composite obtained by RTM process." *Composites: Part A.* 31:151-163.

[17] Uenoya, T. 1995. "Acoustic emission analysis on interfacial fracture of laminated fabric polymer matrix composite." *Journal of Acoustic Emission* 13:S95-S102.

[18] De Groot, P.J., and Wijnen, P.A.M. 1995. "Real-time frequency determination of Acoustic Emission for different fracture mechanisms in carbon/epoxy composites." *Comp. Sci. and Technol* 55:405-412.

[19] Gutkin, R., Green, C.J., Vangrattanachai, S., Pinho, S.T., Robinson, P., and Curtis, P.T. 2011. "On acoustic emission for failure investigation of

CFRP: pattern recognition and peak frequency analyses." *Mechanical Systems and Signal Processing* 25:1393-1407.

[20] Aggelis, D.G., Barkoula, N.M., Matikas, T.E., and Paipetis, A.S. 2012. "Acoustic structural health monitoring of composite materials." *Composites Science and Technology* 72(10):1127-33.

[21] Sause, M., Gribov, A., Unwin, A.R., and Horn, S. 2012. "Pattern recognition approach to identify natural clusters of acoustic emission signals." *Pattern Recognition Letters* 33:17-23.

[22] Doan, D.D., Ramasso, E., Placet, V., Zhang, S., Boubakar, L., and Zerhouni, N. 2015. "An unsupervised pattern recognition approach for AE data originating from fatigue tests on polymer-composite materials." *Mechanical Systems and Signal Processing* 64-65:465–478.

[23] Marec, A., Thomas, J.H., and El Guerjouma, R. 2008. "Damage characterization of polymer-based composite materials: multivariable analysis and wavelet transform for clustering acoustic emission data." *Mechanical Systems and Signal Processing* 22:1441-64.

[24] Li, L., Lomov, S.V., Yan, X. et al. 2014. "Cluster analysis of acoustic emission signals for 2D and 3D woven glass/epoxy composites." *Composites Structures* 116:286-299.

[25] Kostopoulos, V., Loutas, T.H., Kontsos, A., Sotiriadis, G., and Pappas, Y.Z. 2003. "On the identification of the failure mechanisms in oxide/oxide composites using acoustic emission." *NDT&E International* 36(8):571-580.

[26] Kostopoulos, V., Pappas, Y.Z., and Markopoulos, Y.P. 1999. "Fatigue damage accumulation in 3-dimensional SiC/SiC composites." *Journal of the European Ceramic Society* 19(2):207-215.

[27] Kostopoulos, V., Loutas, T., and Dassios, K. 2007. "Fracture behavior and damage mechanisms identification of SiC/glass ceramic composites using AE monitoring." *Composites Science and Technology* 67(7-8):1740-46.

[28] Godin, N., Huguet, S., Gaertner, R., Salmon, L. 2004. "Clustering of acoustic emission signals collected during tensile tests on unidirectional glass/polyester composite using supervised and unsupervised classifiers." *NDT&E International* 37(4):253-264.

[29] Godin, N., Huguet, S., and Gaertner, R. 2005. "Integration of the Kohonen's self-organising map and k-means algorithm for the segmentation of the AE data collected during tensile tests on cross-ply composites." *NDT&E International* 38(4):299-309.

[30] Moevus, M., Rouby, D., Godin, N., R'Mili, M., Reynaud, P., Fantozzi, G., and Fayolle, G. 2008. "Analysis of damage mechanisms and associated acoustic emission in two SiC/(Si-B-C) composites exhibiting different tensile curves." Part I: Damage patterns and acoustic emission activity. *Composite Science and Technology*. 68(6):1258-1265.

[31] Moevus, M., Godin, N., R'Mili, M., Rouby, D., Reynaud, P., Fantozzi, G., Fayolle, G. 2008. "Analysis of damage mechanisms and associated acoustic emission in two SiC/(Si-B-C) composites exhibiting different tensile curves." Part II: Unsupervised acoustic emission data clustering. *Composite Science and Technology*. 68(6):1258-1265.

[32] Barton, T.F., Tuck, D.L., and Wells, D.B. 1996. "The identification of pitting and crevice corrosion spectra in electrochemical noise using an artificial neural network." *ASTM Special Technical Publication* 277: 157–169.

[33] Van Dijck, G., and Van Hulle, M.M. 2011. "Genetic algorithm for informative basis function selection from the wavelet packet decomposition with application to corrosion identification using acoustic emission." *Chemometrics and Intelligent Laboratory Systems* 107:318–332.

[34] Ono, K., and Huang, Q. 1994. "Pattern recognition analysis of acoustic emission signals. Progress in Acoustic Emission VII. The Japanese Society for NDI." 69-78.

[35] Jain, A.K, Murty, M.N., Flynn, P.J. 1999. "Data clustering: a review." *ACM Computing Surveys* 31(3):264-323.

[36] Jain, A.K. 2010. "Data clustering: 50 years beyond K-means." *Pattern Recognition Letters* 31(8):651-666.

[37] Momon, S., Moevus, M., Godin, N., R'Mili, M., Reynaud, P., Fantozzi, G., and Fayolle, G. 2010. "Acoustic emission and lifetime prediction during static fatigue tests on ceramic matrix composite at high temperature under air." *Composites: Part A* 41:913-18.

[38] Maillet, E., Godin, N., R'Mili, M., Reynaud, P., Lamon, J., and Fantozzi, G. 2012. "Analysis of Acoustic Emission energy release during static fatigue tests at intermediate temperatures on Ceramic Matrix Composites: towards rupture time prediction." *Comp Sci Technol* 72:1001-7.

[39] Maillet, E., Godin, N., R'Mili, M., Reynaud, P., Lamon, J., and Fantozzi, G. 2012. "Determination of acoustic emission sources energy and application towards lifetime prediction of ceramic matrix composites." *Mechanical Properties and Performance of Engineering*

Ceramics and Composites VII (Eds D Singh, J Salem, M Halbig and S Mathur), Wiley 33(2):15-25.

[40] Huguet, S., Godin, N., Gaertner, R., Salmon, L., and Villard, D. 2002. "Use of acoustic emission to identify damage modes in glass fiber reinforced polyester." *Comp Sci Technol.* 62:1433-1444.

[41] Godin, N., Huguet, S., and Gaertner, R. 2006. "Influence of hydrolytic ageing on the acoustic emission signatures of damage mechanisms occurring during tensile tests on a polyester composite: Application of a Kohonen's map." *Composite Structures* 72(1):79-85.

[42] Momon, S., Godin, N., Reynaud, P., R'Mili, M., and Fantozzi, G. 2012. "Unsupervised and supervised classification of AE data collected during fatigue test on CMC at high temperature." *Composites Part A* 43:254-260.

[43] Sibil, A., Godin, N., R'Mili, M., Maillet, E., and Fantozzi, G. 2012. "Optimization of acoustic emission data clustering by a genetic algorithm method." *Journal of Nondestructive Evaluation* 31(2):169-180.

[44] Maillet, E., Godin, N., R'Mili, M., Reynaud, P., Fantozzi, G., and Lamon, J. 2014. "Damage monitoring and identification in SiC/SiC minicomposites using combined acousto-ultrasonics and acoustic emission." *Comp: Part A* 57:8-15.

[45] Morizet, N., Godin, N., Tang, J., Maillet, E., Fregonese, M., and Normand, B. 2016. "Classification of acoustic emission signals using wavelets and Random Forests: Application to localized corrosion Original." *Mechanical Systems and Signal Processing* 70–71:1026-1037.

[46] Hsu, N.N., and Breckenridge, F.R. 1981. "Characterization and calibration of acoustic emission sensors." *Materials Evaluation* 39: 60-68.

[47] Morscher, G.N. 2002. "The velocity and attenuation of acoustic emission waves in SiC/SiC composites loaded in tension." *Comp Sci Technol.* 62:1171-1180.

[48] Hamsi, S., Le Duff, A., Simon, L., Plantier, G., Sourice, A., and Feuilloy, M. 2013. "Acoustic emission pattern recognition approach based on hilbert-huang transform for structural health monitoring in polymer-compsoite." *Applied Acoustics* 74:746-757.

[49] Anastassopoulos, A.A., and Philippidis. T.P. 1995. "Clustering methodology for the evaluation of acoustic emission from composites." *Journal of Acoustic Emission* 13(1-2):11-22.

[50] Anastassopoulos, A.A., Philippidis, T.P, and Paipetis, S.A. 1996. "Failure mechanism identification in composite materials by means of acoustic emission: Is it possible?" *Non Destructive Testing*, Eds Van Hemelrijck and Anastassopoulos, A.A. Balkema, Rotterdam: 143-149.

[51] Forgy, E. 1965. "Cluster analysis of multivariate data: Efficiency vs. interpretability of classifications." *Biometrics* 21:768-69.

[52] Likas, A., Vlassis N., and Verbeek, J. 2003. "The global k-means clustering algorithm." *Pattern Recognition* 36(2):451-461.

[53] Dunn, J.C. 1973. "A fuzzy relative of the ISODATA process and its use in detecting compact well-separated clusters." *Journal of Cybernetics* 3:32-57.

[54] Halkidi, M., Batistakis, Y., and Vazirgiannis, M. 2001. "On clustering validation techniques." *Journal of Intelligent Information Systems* 17(2):107-145.

[55] Halkidi, M., Vazirgiannis, M. 2008. "A density-based cluster validity approach using multi-representatives." *Pattern Recognition Letters* 29(6):773-786.

[56] Davies, D.L., and Bouldin, D.W. 1979 "A cluster separation measure." *IEEE Transactions on Pattern Analysis and Machine Intelligence* 1:224-227.

[57] Rousseeuw, P.J. 1987. "Silhouettes: a graphical aid to the interpretation and validation of cluster analysis." *Journal of Computational and Applied Mathematics* 20:53-65.

[58] Keller, J.M., Gray, M.R., and Given, J.A. Jr. 1985. "A fuzzy K-nearest neighbor algorithm. IEEE Transactions on Systems." *Man and Cybernetics* 15(4):580-585.

[59] K., Hattori, M., Takahashi, 1999. "A new nearest-neighbor rule in the pattern classification problem." *Pattern Recognition* 32:425-432.

[60] Kohonen, T. 1974. "Self-organization and associative memory." *Springer series in Information Sciences* 8, Springer-Verlag, Berlin (1984).

[61] Breiman, L. 2001. "Random forests." *Machine Learning* 45:5–32.

[62] Yang, J., Dumont, G. A. 1992. "Classification of mixed acoustic emission signals via neural networks." *Applications of Artificial Neural Networks III*, Proc. SPIE1709:849-854.

[63] Shaira, M., Guy, P., Godin, N., Vanel, L., and Courbon, J. 2010. "Monitoring of martensitic transformation in austenitic stainless steel 304L by Eddy currents." *Research in Nondestructive Evaluation,* 21(2):112-126.

[64] Carrillo, L., and Ortin, J. 1997. "Avalanches in the growth of stress-induced martensites." *Phys. Rev.* B 56:11508–11517.

[65] Sibil, A, Erauw, J.P., Cambier, F.M., R'Mili, M., Godin, N., and Fantozzi, G. 2009. "Study of damage of high zirconia fused-cast refractories by measurement of Young's modulus" *Material Science and Engineering A* 521-22:221-223.

[66] Sibil A., Douillard, T., C. Cayron, N. Godin, M. R'mili, G. Fantozzi. (2011) Microcracking of high zirconia refractories after t→m phase transition during cooling: An EBSD study. *Journal of the European Ceramic Society*, 31 (9): 1525-1531.

[67] Fantozzi, G., Reynaud, P. "Mechanical behaviour of SiC fiber reinforced ceramic matrix composites. Chapter in Comprehensive hard materials." V.K. Sarin, Elsevier 345-366.

[68] Morscher, G., and Godin, N. 2014. "Use of acoustic emission for ceramic matrix composites." *Ceramics Matrix Composites*, Bansal NP and Lamon J., Willey: 520-547.

[69] Racle, E., Godin, N., Reynaud, P., R'Mili, M., Fantozzi, G., Marcin, L., Herb, V., Bouillon, F., and Kaminski, M. 2014. "Indicator for the damage evolution of a SiC/ (Si-B-C) composite subjected to cyclic and static loading at 450°C." *Mechanical Properties and Performance of Engineering Ceramics and Composites* Willey 15-26.

[70] Godin, N., Reynaud, P., R'Mili, M., and Fantozzi, G. 2016. "Mechanical behaviour of Ceramic Matrix Composite and lifetime prediction by Acoustic Emission." *Ceramic Matrix Composites* Edited by: J. Paulo Davim published by De Gruyter: 1-37.

[71] Kim, Y., Fregonese, M., Mazille, H., Feron, D., and Santarini, G. 2006. "Study of oxygen reduction on stainless steel surfaces and its contribution to acoustic emission recorded during corrosion processes." *Corrosion Science* 48:3945–3959.

[72] Kim, Y., Fregonese, M., Mazille, H., Feron, D., and Santarini, G. 2003. "Ability of acoustic emission technique for detection and monitoring of crevice corrosion on 304l austenitic stainless steel." *NDT&E International* 36:553–562.

[73] Sause, M., Horn, S. 2010. "Simulation of acoustic emission in planar carbon fiber reinforced plastic specimens." *Journal of Nondestructive Evaluation* 29:123-142.

[74] Sause, M., Hamstad, M., and Horn, S. 2013. "Finite element modelling of lamb wave propagation in anisotropic hybrid materials." *Composites part B* 249-257.

[75] Aggelis, D.G., and Matikas, T. 2012. "Effect of plate wave dispersion on the acoustic emission parameters in metals." *Computers and Structures* 98-9:17-22.

[76] Gorman, M.R. 1991. "Plate wave acoustic emission." *Journal of the Acoustical Society of America* 90(1):358-364.

[77] Gorman, M.R., and Ziola, S. 1991. "Plate waves produced by transfert matrix cracking." *Ultrasonics* 29:245-251.

[78] Carpinteri, A., Lacidogna, G., Accornero, F., Mpalaskas, A.C, Matikas, T.E., and Aggelis, D.G. 2013. "Influence of damage in the acoustic emission parameters." *Cement and concrete composites* 44:9-16.

[79] Goujon, L., Baboux, J.C. 2003. "Behaviour of acoustic emission sensors using broadband calibration techniques." *Measurement Science and Technology* 14:903-908.

[80] Monnier, T., Dia, S., Godin, N., and Zhang, F. 2012. "Primary calibration of AE sensors by the method of reciprocity. Theoretical and experimental considerations." *Journal of Acoustic Emission* 30:152-166.

[81] Wu, K.L., Yang, M.S., and Hsieh, J.N. 2009. "Robust cluster validity indexes." *Pattern recognition* 42:2541-2550.

[82] Maillet, E. 2012. "Identification des mecanismes d'endommagement et prévision de la durée de vie des composites à matrice céramique par émission acoustique." *PhD Thesis INSA de Lyon* (Identification of damage mechanisms and lifetime prediction for ceramic matrix composites with acoustic emission, PhD Thesis INSA de Lyon).

[83] Bouveyron, C., and Girad, S. 2009. "Robust supervised classification with misture models: learning from data with uncertain labels." *Pattern Recognition* 42:2649-2658.

[84] Maillet, E., Godin, N., R'Mili, M., Reynaud, P., Fantozzi, G., and Lamon, J. 2014. "Real-time evaluation of energy attenuation: A novel approach to acoustic emission analysis for damage monitoring of ceramic matrix composites." *Journal of the European Ceramic Society* 34(7):1673-1679.

[85] Moevus, M. 2007. "Mécanismes d'endommagement, émission acoustique et durées de vie en fatigue statique du composite SiCf/[Si-B-C] aux températures intermédiaires (<800°C)" PhD Thesis INSA de Lyon, 2007 (Identification of damage mechanisms and lifetime prediction for ceramic matrix composites SiCf/[Si-B-C] at intermediate temperature (<800°C)), PhD Thesis INSA de Lyon).

[86] Morscher, G.N. 1999. "Modal acoustic emission of damage accumulation in a woven SiC/SiC composite." *Compos Sci Technol* 59: 687-97.

[87] Le Gall, T. 2016. "Simulation de l'émission acoustique: aide à l'identification de la signature acoustique des mécanismes d'endommagement." *PhD Thesis INSA de Lyon*. (Simulation of Acoustic Emission: a tool for the identification of the acoustic signature of damage mechanisms, PhD Thesis INSA de Lyon).

[88] Gorman, M.R. 1998. "Some corrections between AE testing of large structures and small samples." *Nondestructive testing and evaluation* 14(1):89-104.

In: Focus on Acoustic Emission Research ISBN: 978-1-53611-848-3
Editors: C. Barile and G. Pappalettera

Chapter 2

FATIGUE DAMAGE PROPAGATION IN AISI 430 F/1: A MONITORING APPROACH BASED ON ACOUSTIC EMISSION AND INFRARED THERMOGRAPHY

***Claudia Barile*[*], *Caterina Casavola*,**
Giovanni Pappalettera* and *Carmine Pappalettere
Dipartimento di Meccanica, Matematica e Management,
Politecnico di Bari, Bari, Italy

ABSTRACT

This paper presents a review of the results obtained by studying crack propagation in AISI 430F/1 by using Acoustic Emission (AE) and Infrared Thermography (IT) at the same time. Both the techniques share the capability to allow continuous monitoring during the entire life of the specimen; in fact, differently by other non-destructive techniques (i.e., ultrasound), the acquisition does not require scanning. To implement AE technique two sensors were placed on the surface of the specimen to detect the elastic waves generated by the activation of the defects inside the material subjected to stress. Several features of the recorded acoustic signal were analyzed in order to correlate their variation with the crack propagation process. Data were analyzed also applying different kinds of

[*] Corresponding Author: claudia.barile@poliba.it.

post-processing filters in order to isolate, for example, only events occurring near the slots of the sample or characterized by a given level of amplitude. Appropriate filtering of the cumulative graph of the hits was observed to display a behavior that can be correlated with different stages of propagation of the crack. Interestingly the amplitude analysis of the emitted signal shows the presence of the principal bands of amplitude. Furthermore a thermocamera was set-up to get information connected with temperature variation of the specimen under test. Local rise in the temperature about the crack tip was used to follow crack evolution during the test. Finally, a comparison between the two techniques was carried out aiming to assess differences between the two approaches in following the damage evolution.

Keywords: fatigue damage propagation, AISI 430F/1

1. Introduction

The most part of the mechanical structures is subjected to cyclic loads; as a direct consequence, the understanding of the fatigue behavior is a hot topic in mechanical engineering; crack propagation, in particular, can be identified to be the major cause of failure [1 - 2].

Fatigue failure is the tendency of a material to fracture by means of progressive brittle cracking under repeated alternating or cyclic stresses whose intensity is significantly below the normal strength [3]. Even if the fracture is of a brittle type, it requires some time to propagate, depending on both the intensity and frequency of the stress cycles. However, usually, there is no evident warning during crack propagation so that periodic inspection of the structure must be properly planned if sudden collapse has to be avoided. Frequency of these inspections can be optimized based on expertise or the presence of specifically standard but, however, it would be very useful to have the possibility to monitor continuously the structure without altering the functionality of the structure itself.

In this paper the efficiency of Acoustic Emission (AE) and Infrared Thermography (IT) in monitoring damage evolution in metallic specimens is critically analyzed. The importance of crack monitoring becomes crucial in case of critical parts since unexpected failures could be dangerous not only for the structures but also for people. In this framework, AE technique seems to be a unique method for evaluating the internal status of materials by the detection of the internal stress waves [4 - 5] connected with defect activity [6] and its use has been successfully investigated in many different applications including thermal loading [7 - 10] and mechanical loading [11 - 14]. IT is a powerful non-destructive technique (NDT) whose adoption is consolidated in many fields including damage modes analysis [15 - 16]; acoustic emission and thermal emission can be viewed as two different energy dissipation channels active during cyclic loading conditions. IT is a full-field technique so that it directly allows damage localization. AE can allow localization by placing a net of sensors on the sample and by triangulation of the arrival time of the wave on each detector. The two approaches also share the capability to allow continuous monitoring of the structure without the necessity to stop the test or the proper working of the part [17, 18].

Scientific literature present several attempts to combine both the approaches to follow damage evolution. Most of them are devoted to composites studied both in static and fatigue conditions. In [19] ceramic composites under cyclic loading are analyzed, in [20] damage evolution in Nextel 312/Blackglas™ is studied by detecting AE during tensile test and recording change in surface temperature at the same time. Also the rate of damage evolution was studied in 2D carbon-fiber/SiC-matrix by [21] and in glass fiber reinforced thermoplastic polypropylene in [22]. In spite of this, almost no data are present in literature concerning with the application of AE + IT to metals. The approach was only tested by [23] for aluminum alloy and by [16] for Ti grade 5. In this study the combined AE + IT approach is applied to stainless steel; expected results cannot be simply inferred by tests on other metallic samples due to the different thermal conductivity and critical shear stress, frictional stress and surface energy involved. In addition, limited to AE, it should be observed that several papers deal with their application in fatigue of metals in general [24] and also with steel in particular [25, 26]. However, they do not present a complete analysis of AE parameters and in particular of amplitude and location of acoustic source as it is here discussed. Finally, a comparison among the data both from AE and IT techniques is here provided and indications about their capability in following damage evolution in stainless steel samples under uniaxial fatigue loading is discussed.

2. Materials and Methods

Experimental tests were carried out on eight AISI 430F/1 specimens. This is a stainless steel having good machinability especially in the operation of metal cutting. Specimens have a rectangular cross section A = 40x4 mm^2 and a length l = 400 mm. Two notches were cut on each specimen in correspondence of the axial mid-length on both edges symmetrically with respect to the median line. Notches were obtained by laser cutting and have a radius of curvature r = 5 mm. The presence of notches allows to introduce, in the sample, a region of stress concentration where crack nucleation and propagation is promoted. The stress concentration for the said geometry was computed by FEM analysis and it was found to be $k_t = 3$. Figure 1 it is possible to see the maximum stress concentration for a specimen subjected to a nominal unitary stress. In this way, the computed stress at crack tip corresponds, by definition, to the stress intensity factor.

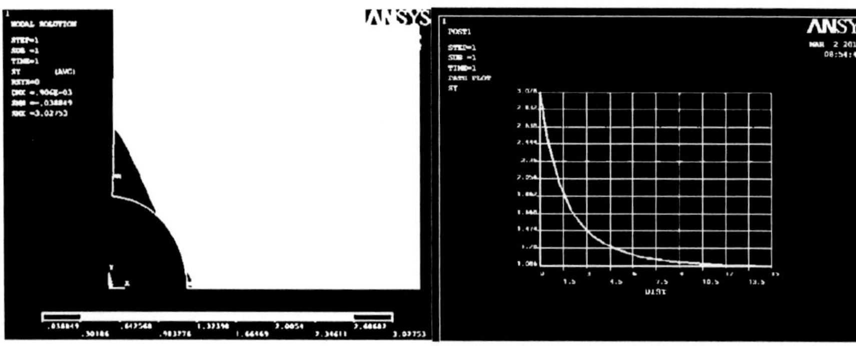

Figure 1. Results obtained by numerical model for stress intensity factor calculation.

Fatigue tests were performed by using an Instron servo-hydraulic machine equipped with a 100 kN loading cell. A fully reversible sinusoidal loading was applied to the specimens by using a loading ratio R = -1 and a loading frequency f = 3 Hz. Three different levels of stresses were tested, respectively $\sigma_0 = 91.7$ MPa, $\sigma_0 = 125$ MPa and $\sigma_0 = 150$ MPa. Two strain gages with a 3 mm base length were placed near the slots and their signals were acquired by the StrainSmart 5000 acquisition system. The acquisition frequency was set to 10 Hz. To detect AE two piezoelectric sensors [27, 28], were coupled with the sample by a silicon gel. They were placed 80 mm far apart from each other and symmetrically located with respect to the transverse median line crossing the two notches (Figure 2). Table 1 report the main characteristics of the sensors used in this experiment.

The possibility to record the AE by two different sensors allows 1D localization of the acoustic source along the longitudinal direction of the sample. The localization algorithm is based upon the recording of the arrival time of the acoustic signal on each sensor. If the distance between the two sensors and the speed of the sound in the material are known, it is possible to calculate the position of the event.

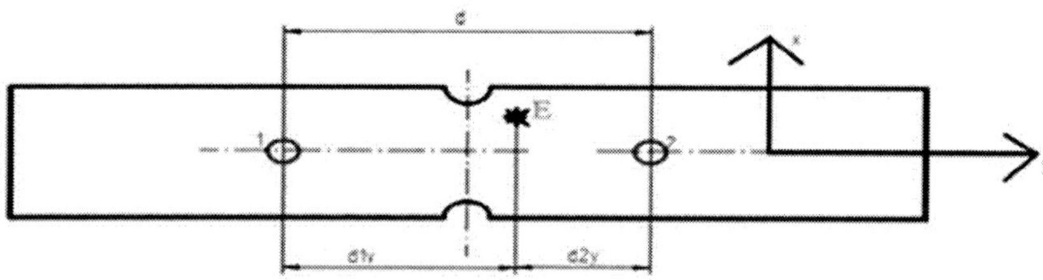

Figure 2. Schematic of the location of the PZT sensors on the sample.

Table 1. Main characteristics of the Mistras Pico sensor used in this work

Characteristics of the Pico Sensor	
Peak Sensitivity	54 dB
Operating Frequency Range	200 – 750 kHz
Resonant Frequency	250 kHz
Temperature Range	-65 to 177 °C
Dimension	5 x 4 mm

The determination of the speed of the sound was preliminary done by performing a Hsu-Nielsen test. Namely a pencil tip was broken at a given position on the sample in order to generate a well-known and repeatable acoustic event, then the speed of the sound for the analyzed material was calculated and it was found to be v = 5500 m/s. Once that v and the distance between the sensors (d_s) are known it is possible to perform linear localization of the AE source by solving the following equation:

$$|d_{1y}-d_{2y}|=v\cdot|T_1-T_2| \quad (1)$$

By knowing that:

$$d_s=d_{1y}+d_{2y} \quad (2)$$

where T_1 and T_2 represent the arrival time of the wave on each sensor while d_{1v} and d_{2v} represent the distances travelled by the wave to reach the sensors.

Acoustic emissions acquisition frequency was chosen in consideration of the duration of the test and set to 100 kHz. While a threshold value of 45 dB was chosen because preliminary tests demonstrated that it efficiently reduces background noise without altering the main features of the signal.

Thermographic images were recorded by means of a NEC H2640 thermocamera with a 640 x 680 pixel sensor matrix. Thermocamera was set to detect a temperature range between - 20 °C and + 120 °C. In order to increase the emissivity of the sample and avoid reflections, a black matt spray paint ($\varepsilon = 0.97$) was applied on the front surface. The thermocamera was placed in front of the specimen and perpendicularly to it. The working distance was set to d = 40 cm in order to get the minimum field of view obtainable by the equipment. Acquisition frequency was set to f = 30 Hz that is the maximum achievable for the instrument.

The overall experimental set-up is shown in Figure 3 while in Figure 4 the geometry and a picture of the specimen are displayed. Figure 5 shows a detail of the positioning of both the PZT sensors and the strain gages.

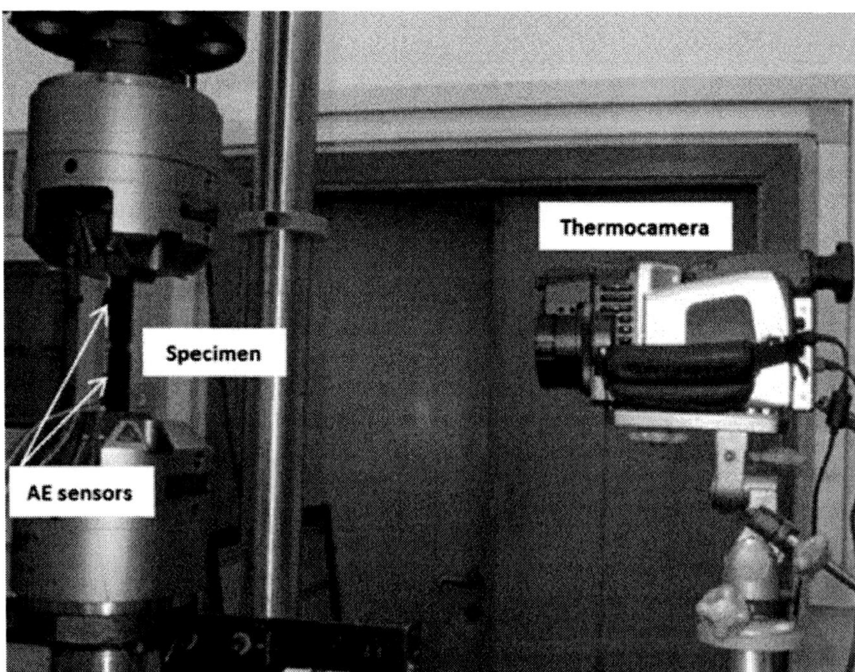

Figure 3. Experimental set-up; the main elements are the loading machine, the piezoelectric sensors and the thermocamera.

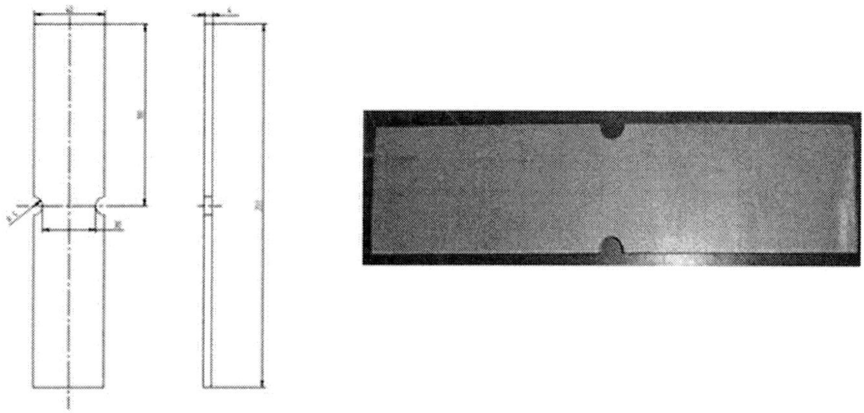

Figure 4. (left) Sketch and dimensions of the specimens; (right) picture of the specimen.

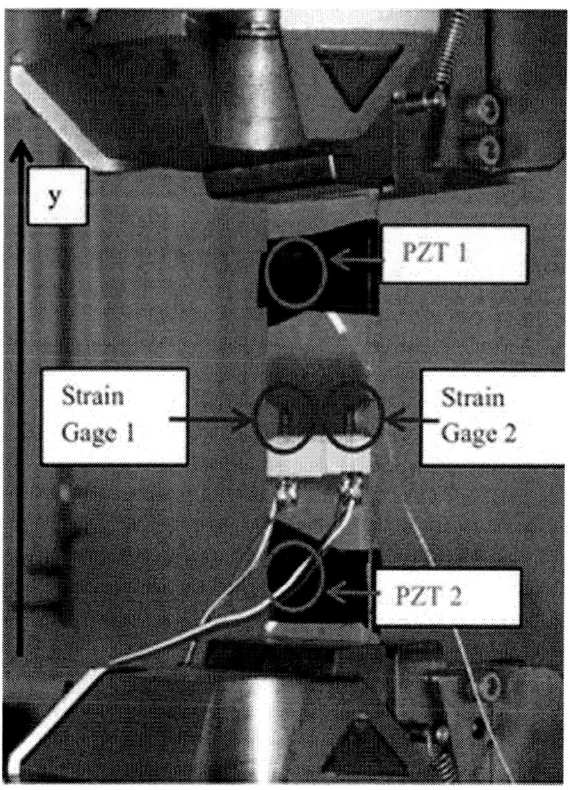

Figure 5. Picture of the set-up with the indication of the position of the acoustic piezoelectric sensor and the strain gages.

3. Results and Discussion

The Figure 6 shows an exemplary thermal map recorded during the test. In particular it can be seen that a ROI is identified on the image, indicated by the black box. In the ROI area the temperature recorded was maximum in correspondence of the crack tip. Figure 6 also shows a line which indicating the cross section adopted for temperature profile calculation.

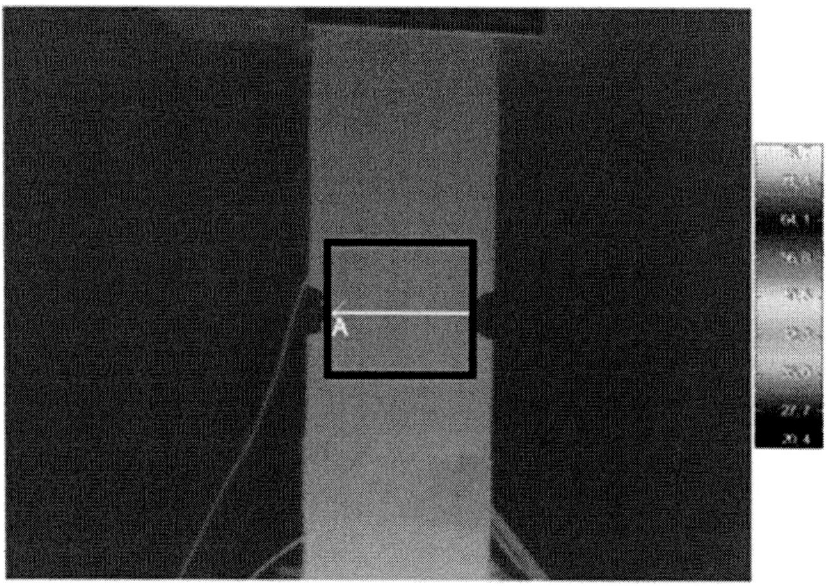

Figure 6. Thermal image: the ROI and the cross section A are also indicated.

By analyzing the temperature profile along the cross section A it was possible to observe, from a given moment in the test on, the presence of a maximum of temperature. The maximum corresponds to the position of the crack tip. Furthermore, it can be observed that this peak moved along the direction of crack propagation increasing its T value as the crack size increased (Figure 7).

This behavior can also be seen by observing a sequence of three thermal images recorded at different stages during the fatigue test (Figure 8). From left to right it is possible to see that the crack tip appears more and more observable as the test goes on. In the left image, the temperature increment, with respect to the surrounding area, is limited but ΔT tends to increase during the test. Capability to precisely identify the position of the crack tip is evident by observing the figure and it is a natural consequence of adopting a high-resolution full-field technique. It is also evident that by analyzing the sequence of thermal maps it is quite easy to extract information about the damage evolution path. Figure 8, for example, shows that crack propagates from the right to left almost along the line connecting the centers of the two notches. However, it should be underlined that temperature increment is not observable in all stages of the test. In the tested sample, it starts to be evident only when the crack has travelled about half of the sample. This means that all the above observations should be considered valid only starting from an advanced stage of the overall fatigue test.

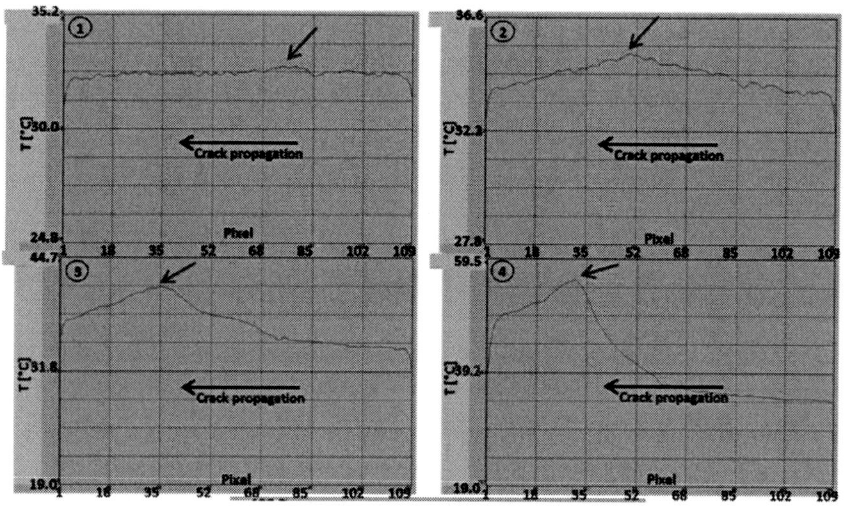

Figure 7. Thermal profile recorded during fatigue crack propagation.

Figure 8. Sequence of thermographic images indicating the position of the crack tip at different stages of the fatigue test.

This issue can be better understood by plotting the evolution of the temperature at crack tip as a function of the normalized number of cycles, that is to say the number of cycles normalized to the total number of cycles of the test (Figure 9). It is evident that the temperature stays almost stable for the greater part of the test while an abrupt increase appears in the last stage.

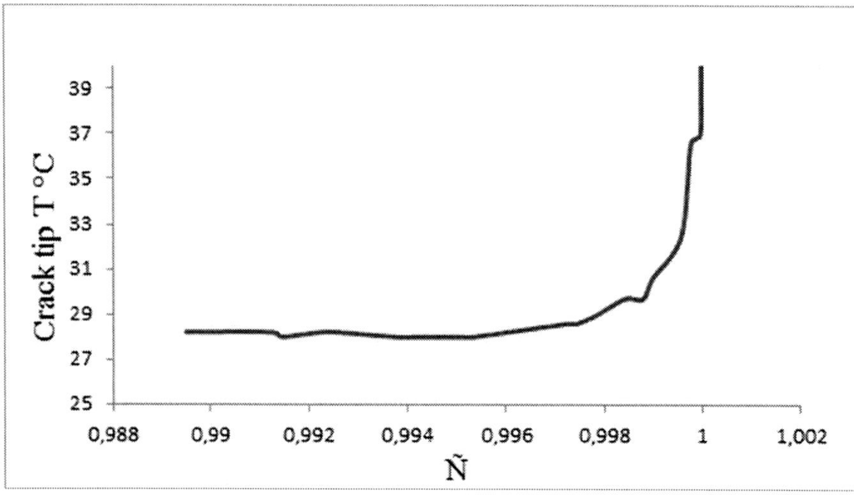

Figure 9. Thermal behavior of the temperature at the crack tip as a function of the normalized number of cycles.

By studying the thermograms it is also possible to analyze how the crack size increases during the test. *Figure* 10, shows that crack size does not follow a smooth trend but a step-like behavior.

By plotting the crack tip temperature as a function of the crack size, it is possible to observe that temperature increases slowly until the crack reaches the dimension of about a = 20 mm (Figure 11). From that point on the temperature increases much more rapidly. This behavior was found in all the test samples and it is a concern that the IT technique can hardly be used to detect with great advance the impending failure.

In Figure 12 the slope of the F vs ε is plotted as a function of the normalized number of cycles that is to say the current number of cycles divided by the number of cycles to rupture. Data are referred to the two strain gages (es1 and es2) applied on a specimen tested at F = ± 15 kN, however similar behavior has been observed for all the tested specimens.

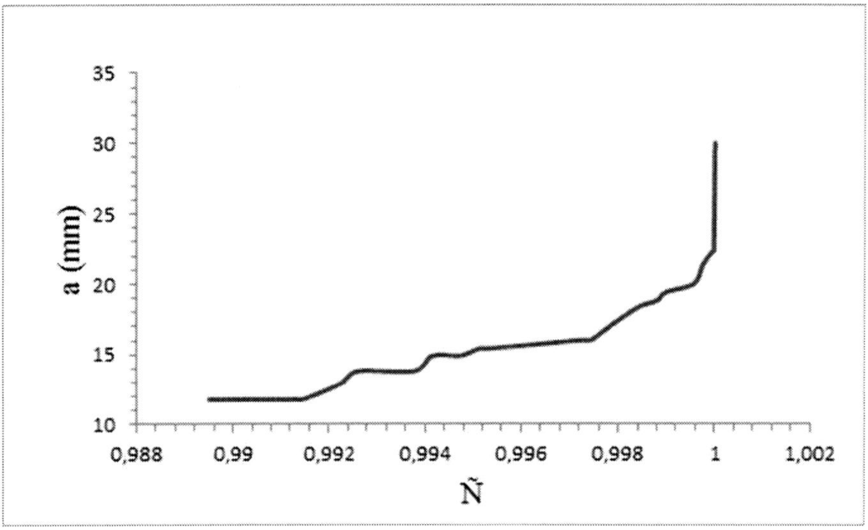

Figure 10. Plot of the crack size as a function of the normalized number of cycles.

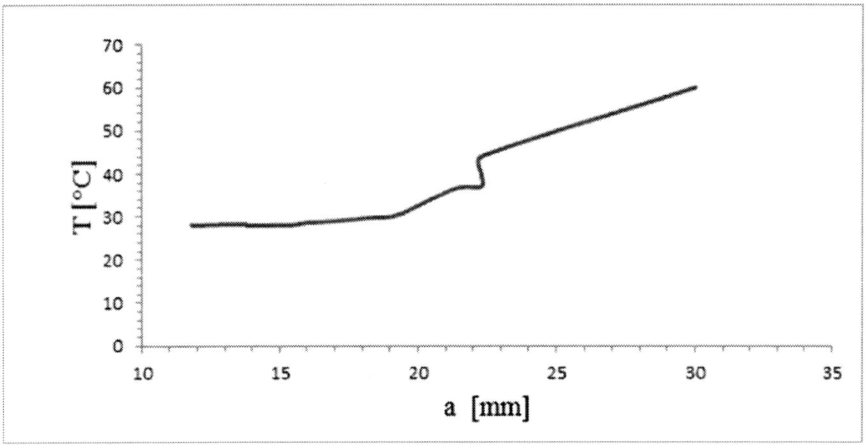

Figure 11. Plot of the temperature at the crack tip as a function of the crack tip position.

A first overall information that can be obtained by analyzing AE data is connected with the amplitude distribution of the recorded events. In Figure 13 3D graph, Events vs Amplitude vs Y Position is represented. It results that highest amplitude events are located in the region close to the notches; this gives a first indication about the fact that high amplitude events are someway connected with damage evolution. However, in this case, detection is efficient also in the first stages of the test.

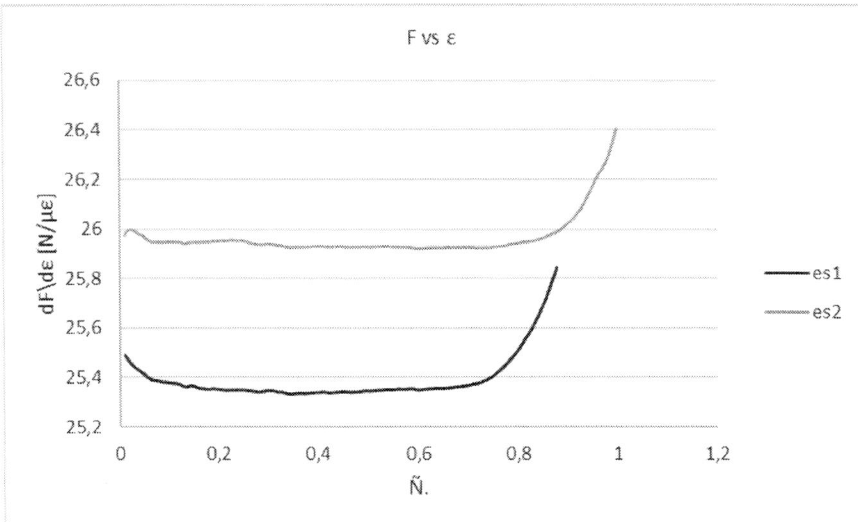

Figure 12. dF/dε vs Ñ curve. Plot was obtained for a specimen at ± 15kN.

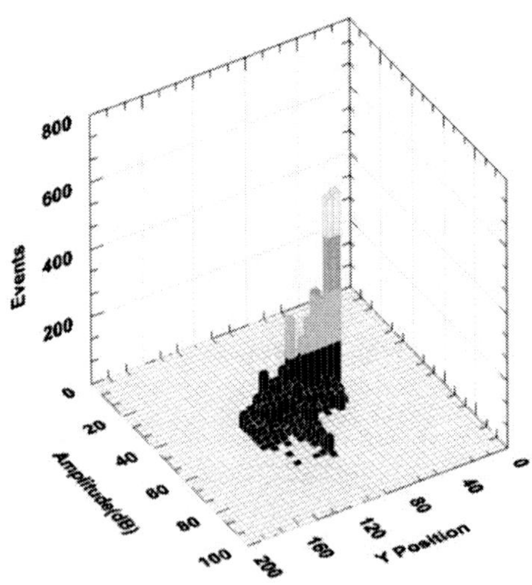

Figure 13. Histogram plot representing the distribution of the recorded acoustic event at a given position along the longitudinal line and with a given amplitude.

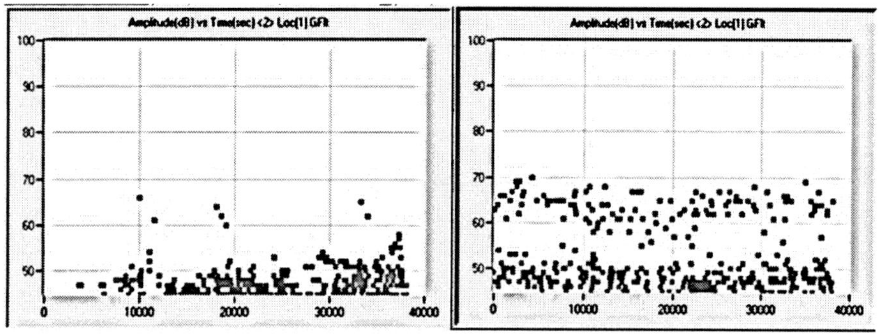

Figure 14. Scatter plot of the amplitude of events localized into two control volumes: far and near the notches.

The preeminence of acoustic activity in the area surrounding the slots can be also pointed out by proceeding with a different data analysis based on the volume density of acoustic emission. To do these, two reference volumes were identified; the dimensions of the volumes are 10 mm x 10 mm x 4 mm and they were ideally placed respectively close to the notches and far from them. The control volume near to the notches stays at the center of the specimen centered along the axis connecting the two notches. The control volume far from the notches stays at 32 mm from the axis connecting the two slots. Based on the triangulation of the arrival time on the two sensors it is possible to separate the contribution to the acoustic emission associated to each reference volume. By doing this kind of analysis, interesting findings can be observed as reported Figure 14. The figure shows the scatter plot of the amplitude of the events recorded during the fatigue test. It is interesting to observe how this distribution of amplitude is different in the two control volumes. In fact, while in the region far from the notch it can be observed an average amplitude of about 50 dB, the situation changes in the region near the notches. In this case a second band appears in the scatter plot, this band is connected with the occurrence of higher amplitude events, averaging about 65 dB. The appearance of this band can be directly connected with crack related events. An amplitude spectrum in steel made up of two different bands associable to two different sources was similarly observed in [26]; however, in the cited paper the higher emission source was only activated and observable starting from a given time along the test, that is to say in correspondence of a given critical stress intensity factor. In our experiment, instead, higher amplitude events are present from the beginning of the test and they are strictly connected with the area around the notches. This is an important finding because it indicates that it is possible, by performing an amplitude analysis, to distinguish and locate different mechanisms of damage evolution in AISI 430F/1 subjected to fatigue loading.

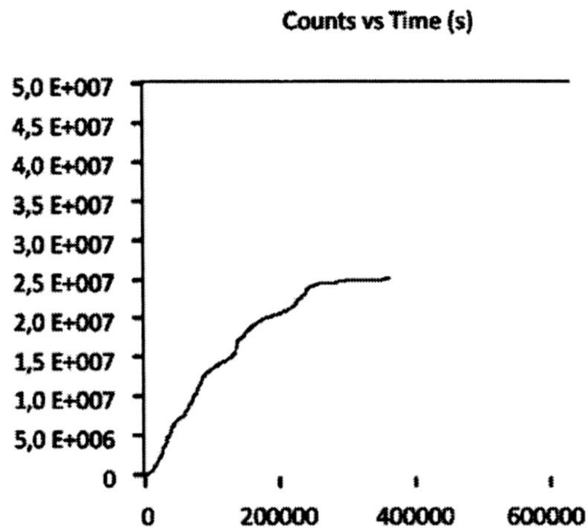

Figure 15. Representative plot of the cumulate number of counts recorded during the test.

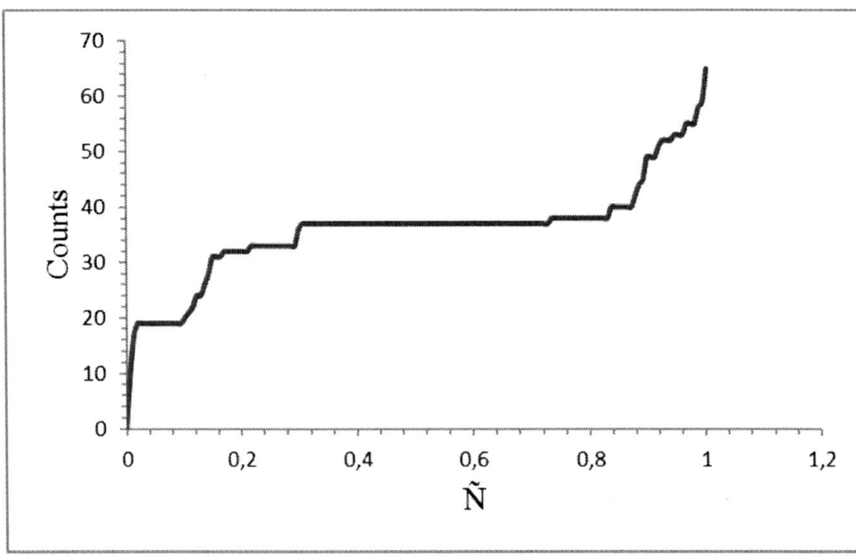

Figure 16. Counts vs Normalized number of cycles. Counts are filtered so that only events with amplitude higher than 60 dB were taken into account.

In Figure 15 a representative plot of the counts progressively recorded on a given PZT sensor is presented. Two main features can be observed. In the first part of the test the number of counts increases rapidly and almost constantly. In the second part of the test the number of counts remains stable until the end of the test and the rupture of the specimen. This behavior differs from analogous analysis performed, for example, in [13] and in [25]. This difference could be due to the fact that, in the final stage of the test, a great number of events occurs; this can mask several features more strictly connected with damage propagation. In order to extract this information, and based upon the observations reported above it was decided to apply, to the data, a high-pass amplitude filter having a threshold of 60 dB. The result of this operation is shown in Figure 16.

This confirms predictive capability of the acoustic emission provided that proper data analysis is carried out in order to extract information relevant to follow damage propagation. This aspect can also be observed from another point of view connected with location of the acoustic events. In fact, if the two above defined control volumes are considered, the evolution of the density of the events is that shown in Figure 17. It is easily observable that only by considering events in the proper area it is possible to extract information connected with damage propagation and impending failure.

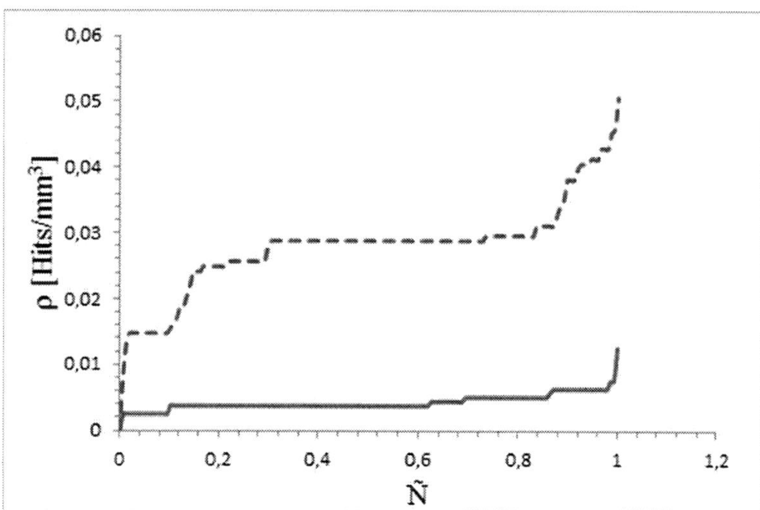

Figure 17. Trend of density of hits recorded in two different regions, close and far from the notch, as a function of the normalized number of cycles. Solid line is referred to a region far from the slots. Dashed line is referred to a region near the slots.

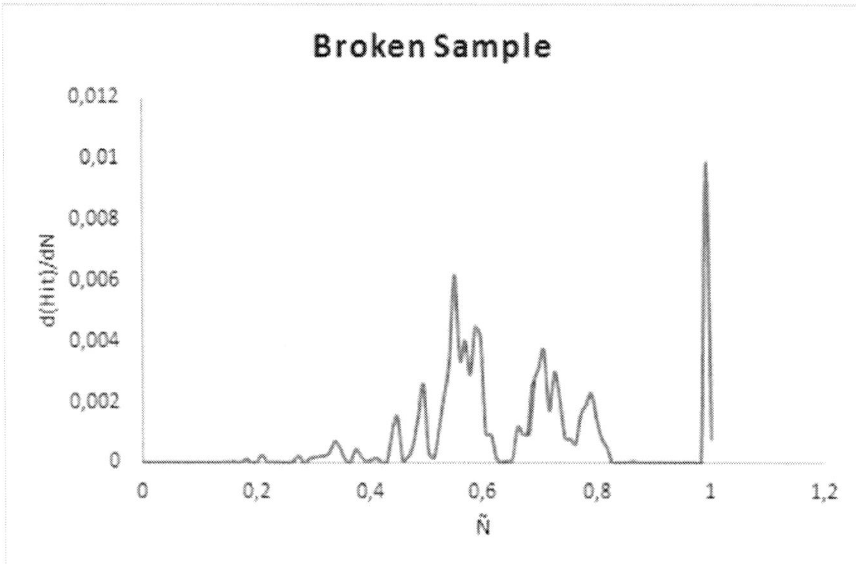

Figure 18. Trend of d(Hits/dN) vs normalized number of cycles (~N): broken sample.

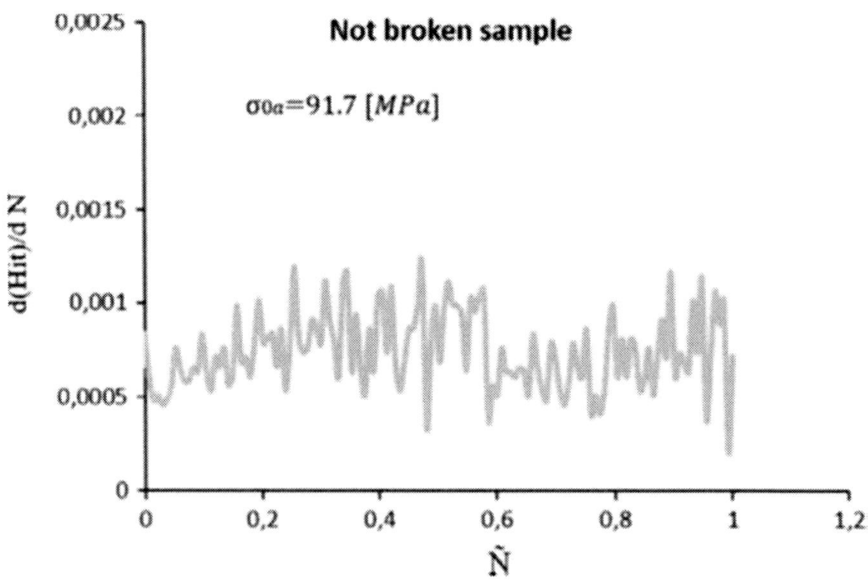

Figure 19. Trend d(Hits/dN) vs normalized number of cycles (~N): not broken sample.

Another way to analyze acoustic activity is presented in Figure 18 and Figure 19. In these graphs the derivatives of the number of hits with respect to the normalized number of cycles (dHits/dÑ) are presented.

Figure 18 is representative of samples that break at the end of the test. It displays the characteristic feature of a sharp peak anticipating the moment of rupture. This peak does not appear in the samples that do not break (Figure 19). Figure 18 also displays that some other peaks of lower amplitude corresponding to a sudden increment of the acoustic activity. The presence of that peak can be explained in terms of microcrack development as a result of the movement of dislocations [13]. These graphs show features that require more investigation but surely add important information that can be inferred by live monitoring of the state of material.

CONCLUSION

In this paper fatigue tests on AISI 430F/1 stainless steel were performed. The specimens during the test were continuously monitored by using a hybrid approach based on two different non-destructive techniques: acoustic emission analysis and infrared thermography.

AE technique was found to be much more efficient in following damage propagation during the whole test allowing also to distinguish crack initiation stage from stable and unstable crack propagation stage. However, this is possible only if proper filtering of the data is applied because several features are masked by the overall activity of the sample. Moreover, in the 1D configuration adopted for the two AE sensors only the longitudinal coordinate of the event is measurable so that location task can be only partially addressed. Location is much more efficient, being an intrinsic aspect of the full-field techniques, if infrared thermography is adopted. By this technique crack tip can be observed and its motion can be monitored with very high accuracy. However, this is not possible during the whole test because temperature increment in the initial part of the test is not high enough to be detected by the thermocamera.

IT becomes efficient to localize and to monitor crack evolution only when crack dimensions become relevant; only at that point the rise of temperature around the crack tip is high enough to allow reliable detection of the crack position and crack path identification. In that situation, however, localization appears to be very precise because the technique provides 2D information with a very good resolution. Combination of both the techniques needs more investigations, since it appears promising as a powerful non-destructive tool for fatigue crack monitoring. Overall the two approaches appear to be complementary in providing information about damage evolution so that a combined approach appears to be a more powerful method for fatigue analysis.

REFERENCES

[1] Dowling, N. E. 2012. "Mechanical Behavior of Materials, Engineering Methods for Deformation, Fracture, and Fatigue". *Pearson.*

[2] UNI 3964. "Prove di fatica a temperatura ambiente".

[3] Milella, P. P. 2013. "Fatigue and Corrosion in Metal". *Springer-Verlag* Italia.

[4] Miller, R. K., and McIntire, P. 1987. "Nondestructive testing handbook" 5, II ed. *American Society for Nondestructive Testing*, Columbus.

[5] Drouillard, T. F., and Glenn, T. G. 1985. "Production acoustic emission testing of braze joint". *Journal of Acoustic Emission* 1:81-85.

[6] Roberts, T. M., and Talebzadeh, M. 2003. "Acoustic emission monitoring of fatigue crack propagation". *Journal of Construction Steel Research* 59:695-712.

[7] Barile, C., Casavola, C., Pappalettera, G., and Pappalettere, C. 2013. "Feasibility of Local Stress Relaxation by Laser Annealing and X-Ray Measurement". *Strain* 49:393-398.

[8] Barile, C., Casavola, C., Pappalettera, G., and Pappalettere, C. 2012. "Preliminary Analysis for a New Approach to Relieve Residual Stresses by Laser Heating". *Conference Proceedings of 11th IMEKO TC15 YSESM* 77-82.

[9] Barile, C., Casavola, C., Pappalettera, G., and Pappalettere, C. 2014. "Acoustic emission analysis of aluminum specimen subjected to laser annealing". *Conference Proceedings of the Society of Experimental Mechanics* 8:309-315.

[10] Barile, C., Casavola, C., Pappalettera, G., and Pappalettere, C. 2013. "Discussion on the thermal field produced by laser annealing for the residual stress relaxation". *Conference Proceedings of New Trends in Fatigue and Fracture* 13, 22-30.

[11] Barile, C., Casavola, C., Pappalettera, G., and Pappalettere, C. 2015. "Hybrid thermography and acoustic emission testing of fatigue crack propagation in Aluminum Samples". *Conference Proceedings of the Society of Experimental Mechanics* 66 (5):247-252.

[12] Nemati, N., Metrovich, B., and Nanni, A. 2015. "Acoustic emission assesment of through-thickness fatigue crack growth in steel members". *Advancing in Structural Engineering* 16:269-282.

[13] Vlasic, F., Nohal, L., Mazal, P., and Liskutin, P. 2014. "Study of high-cycle fatigue behaviour of titanium alloy using acoustic emission method". *Conference Proceedings Metal* 1423-1430.

[14] Wu, X., Botten, R., Hu, D., and Loretto, M.H. 2001. "Cracking of TiAl-based alloys and the effect of cracking on fatigue life". *International Journal of Materials Processing Technology* 1:269-274.

[15] Naderi, M., Kahirdeh, A., and Khonsari, M. M. 2012. "Dissipated thermal energy and damage evolution of Glass/Epoxy using infrared thermography and acoustic emission". *Composites-Part B-Engineering* 43:1613-1620.

[16] Barile, C., Casavola, C., Pappalettera, G., and Pappalettere, C. 2015. "Fatigue Damage Monitoring by means of Acoustic Emission and Thermography in Ti grade 5 specimens". *Procedia Engineering* 114:487-492.

[17] Barile, C., Casavola, C., Pappalettera, G., and Pappalettere, C. 2015. "Considerations on acoustic emissions in Ti Grade 5 during fatigue test". *Procedia Engineering* 109:320-326.

[18] Geng, R. 2006. "Modern acoustic emission technique and its application in aviation industry". *Ultrasonics* 44:e1025-e1029.

[19] Kordatos, E. Z., Aggelis, D. G., Dassios, K. G., and Matikas, T. E. 2013. "In-Situ Monitoring of Damage Evolution in Glass Matrix Composites during Cyclic Loading using Nondestructive Techniques". *Applied Composite Materials* 20(5):961–973.

[20] Kim, J., and Liaw, K. P. 2005. "Monitoring tensile damage evolution in Nextel 312/Blackglas™ composites". *Materials Science & Engineering A* 409(1-2):302-308.

[21] Chidong, L., Laifei, C., Xingang, L., Bin, L., and Jun, Z. 2008. "Damage evolution and real-time non-destructive evaluation of 2D carbon-fiber/SiC-matrix composites under fatigue loading". *Materials Letters* 62(24):3922-3924.

[22] Kocsis, K. J., and Kozma, F. Z. 1997. "Failure mode and damage zone development in a GMT-PP by acoustic emission and thermography". *NDT & E International* 30(2):110.

[23] Kordatos, E. Z., Aggelis, D. G., and Matikas, T. E. 2012. "Monitoring mechanical damage in structural materials using complimentary NDE

techniques based on thermography and acoustic emission". *Composites: Part B* 43:2676–2686.

[24] Aggelis, D. G., Kordatos, E. Z., Matikas, T. E. 2011. "Acoustic emission for fatigue damage characterization in metal plates. *Mechanics Research Communications* 38(2):106-110.

[25] Han, Z., Luo, H., Cao, J., Wang, H. 2011. "Acoustic emission during fatigue crack propagation in a micro-alloyed steel and welds". *Materials Science and Engineering: A-Structural Materials: Properties, Microstructures in Processing* 528: 7751-7756.

[26] Sinclair, A. C. E., Connors, D. C., and Formby, C. L. 1977. "Acoustic emission analysis during fatigue crack growth in steel". *Materials Science and Engineering* 28(2):263-273.

[27] Enoki, M., Kishi, T., and Kohara, S. 1983. "Determination of Microcracking Moment Tensor of Quasi-Cleavage Facet by AE Source Characterization". *Progress in Acoustic Emission III, Proceedings of the Eighth International Acoustic Emission Symposium, The Japanese Society for Non-Destructive Inspection* 763-770.

[28] www.mistrasgroup.com.

In: Focus on Acoustic Emission Research
Editors: C. Barile and G. Pappalettera
ISBN: 978-1-53611-848-3

Chapter 3

THE APPLICATION OF ACOUSTIC EMISSION TECHNIQUE IN THE DELAMINATION OF LAMINATED COMPOSITES

Mohamad Fotouhi*[1,*]*, Milad Saeedifar*[2]*,
***Jalal Yousefi*[2] *and Sakineh Fotouhi*[3]**
[1]Advanced Composites Centre for Innovation and Science, University of Bristol, Bristol, UK
[2]Department of Mechanical Engineering, Amirkabir University of Technology, Tehran, Iran
[3]Department of Mechanical Engineering, University of Tabriz, Tabriz, Iran

ABSTRACT

Fiber reinforced plastic composites are being increasingly used in engineering applications due to their high specific stiffness and strength. On the other hand, the incidence of internal defects may considerably alter the stiffness and reduce the strength and lifetime of the composites. The principal mode of failure in laminated composite materials is delamination. It can be caused by manufacturing faults or subsequent

* Corresponding author: Mohamad Fotouhi. E-mail: fotouhi.mohamad@gmail.com.

operational effects such as impact loads, fatigue, etc. It usually occurs within the structure and it is difficult to distinguish during the inspection. Therefore, it is not easy to detect the delamination existence and to characterize the onset and growth of it. Acoustic Emission (AE) is a capable non-destructive method to study the delamination damage. This chapter represents the application of AE method to investigate delamination of laminated composite materials in two major sections:

- Initiation of damage: Various approaches such as AE energy, AE count, AE cumulative energy, and Sentry function are described as appropriate tools to determine interlaminar fracture toughness (G_C) of composite materials. The AE results are also compared with the results of ASTM standard methods. It is shown that the G_C values obtained by the AE-based methods are in a close agreement with the ASTM results.
- Progression of damage: AE technique is utilized to predict delamination evolution curve, and to classify different damage mechanisms occurring during delamination propagation. In order to predict delamination evolution curve, two different AE-based procedures are presented. In the first method, after determining the velocity of the AE waves in the composite specimens, the position of the crack tip can be estimated using two AE sensors. In the second method, the cumulative energy of the AE signals is utilized for localization of the crack tip. A very good agreement between the AE-predicted crack length and the actual crack length verifies the accuracy of the presented procedures. Finally, wavelet packet transform and fuzzy clustering methods are used to classify the dominant damage mechanisms based on their AE signatures. Scanning electron microscopic observations from the damage mechanisms are in a good agreement with the AE-based classification results.

In conclusion, the results indicate that AE has a good performance to detect initiation stage of delamination, determining the delamination crack length, and discriminating the different types of damage mechanisms in the laminated composite structures.

Keywords: Acoustic emission, laminated composites, delamination, damage characterization.

1. Introduction

High performance laminated composites have been used in different engineering applications, such as aerospace, automotive, etc., because of their high specific stiffness and strength. However, probable damage mechanisms may degrade the long-term performance of these materials. Delamination is a particularly dangerous damage for high performance laminated composites that can be caused due to static or repeated cyclic stresses, impact, and so on. This damage reduces the strength/stiffness and consequently limits the life of composite structures [1-3].

Considerable efforts have therefore been done to characterize delamination in order to modify the performance of the structure [4-5]. Initiation and propagation of the crack are two major stages in delamination damage. Fiber/matrix interface micro damage is not observable and is harder to inspect, whereas, crack propagation is usually a macroscopic damage activity that can be regarded as the ultimate form of delamination occurring in the interlaminar zone and can be observed visually. In laminated composite materials, microscopic damage mechanisms are usually fiber breakage, interface decohesion and matrix cracking.

Acoustic emission (AE) is a promising method to monitor the microscopic damage mechanisms in delamination [6-8]. AE is commonly defined as transient elastic waves within a material, generated by the rapid release of energy within a material because of changes in local stress fields. Acoustic emission waves cover a wide frequency distribution ranges from audible frequencies to frequencies in MHz [9-10]. During delamination, AE signals originated from various sources of the microscopic damage mechanisms contain different AE signatures that can be used to find the relationship between the AE parameters and type of damage mechanisms.

This chapter discusses the application of AE method to investigate delamination of laminated composite materials in the initiation and propagation stages. AE energy, AE count, AE cumulative energy, and Sentry function are used instead of ASTM standard methods to detect the damage initiation and to determine interlaminar fracture toughness (G_C) of composite materials. In addition, AE technique is utilized to predict delamination evolution curve and to classify different damage mechanisms occurring during the delamination propagation. The results indicate that AE is a very efficient method for damage characterization of delamination and can lead to a fundamental understanding of the crack behavior.

2. Experimental Procedure

2.1. Materials

The experimental works were carried out on specimens made from epoxy resin reinforced by carbon and glass fibers. The specifications of the specimens are listed in Table 1. The initial crack was formed by inserting a Teflon film with a thickness of 20 μm at mid-plane of the laminated composite during molding. The laminated composite specimens consisted of a rectangular shape and uniform thickness.

Table 1. The specifications of different composite specimens

Specimen	Thickness (mm)	Length (mm)	Width (mm)	Initial crack length (mm)	Layups	Fiber Type
GU-S1	5	150	25	20	$[0]_n$	Glass
GW-S1	5	150	25	20	$[\text{Woven }(0,90)]_n$	Glass
GU-S2	5	190	22	40	$[0]_n$	Glass
GW-S2	5	190	22	40	$[\text{Woven }(0,90)]_n$	Glass
GU-S3	5	180	25	40	$[0]_n$	Glass
GW-S3	5	180	25	40	$[\text{Woven }(0,90)]_n$	Glass
CU-S1	4	175	25	25	$[0]_n$	Carbon

2.2. Test Design and Configuration

Double Cantilever Beam (DCB), Mixed Mode Bending (MMB), and End-Notched Failure (ENF) test apparatus, shown in Figure 1, were used to apply the load to the specimens. In DCB setup, an upward force is applied to split end of the laminate to create Mode I. While in ENF setup, a downward load is applied to the specimen center to create Mode II. MMB is the combination of DCB and ENF. The length of the MMB lever arm can be changed to vary the GI/GII modal ratio values. The tests were carried out at a temperature of 24°C and at a constant displacement rate of 2 mm/min. The load and displacement were continuously measured and the crack length was visually recorded using a digital camera. The tests were iterated three times for each type of the specimens in constant loading conditions to check the repeatability of the results.

2.3. Acoustic Emission

An AE data acquisition system (PAC) PCI-2 with a maximum sampling rate of 40 MHz was used to record the AE signals. PAC R15 resonant-type, broadband, single-crystal piezoelectric transducers were used as AE sensors. The frequency range of the sensors was 100-500 kHz and the gain selector of the preamplifier and the threshold value were set to 40 dB. The surface of the sensor was covered with silicon grease to provide good acoustic coupling between the specimen and the sensor. The test sampling rate was 1 MHz. A pencil lead break procedure was used to calibrate the data acquisition system for each of the specimens. After the calibration step, the AE signals were recorded during the tests.

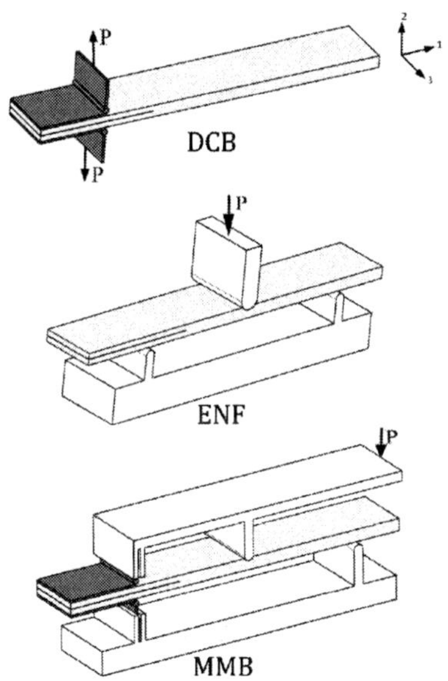

Figure 1. The boundary and loading conditions of the investigated specimens.

3. Results and Discussion

3.1. Damage Initiation

In this section, various approaches such as AE energy, AE count, AE cumulative energy, and Sentry function are used to characterize the initiation stage of delamination, and to determine interlaminar fracture toughness (G_C) of composite materials. The significance of this investigation is due to the fact that initiation is a critical stage in delamination of laminated composites. Knowing the delamination behavior in initiation can lead to an enhanced design with higher strength against crack initiation. In addition to the AE-based procedures, ASTM D5528 [11] standard was also used to detect the initiation stage. The following procedures are presented in ASTM D5528 standard to determine the critical load (P_C) that is required for delamination initiation: a) Non-linearity in the load-displacement diagram (NL), (b) Visual Inspection System (VIS) and (c) The compliance increase (5% max).

Figure 2 depicts values of the P_C that are acquired from the above-mentioned methods for a GW-S1 sample.

Some researchers [2] revealed that the nonlinearity on the load–displacement plot is a practical indication of crack initiation. However, obtaining a deviation from linearity is rather difficult and this method is not completely satisfactory. Besides, visual inspection of delamination is often hard as it occurs beneath composite surfaces in actual applications [13, 14]. In this respect, AE is an effective tool for the initiation detection of delamination. A successful reliability of AE method has been reported for the damage initiation characterization by some researchers [15-17].

Two main AE-based methods for damage initiation detection in composite laminates are presented in the following sections.

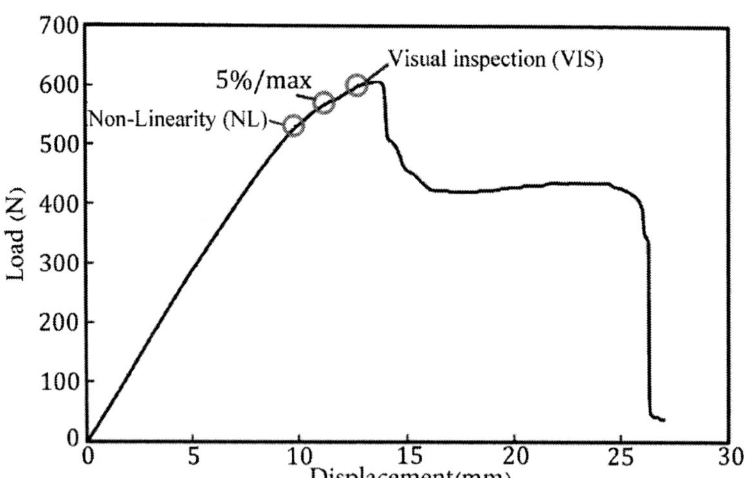

Figure 2. Pc values obtained from the ASTM standard methods for GW-S1 using ENF setup.

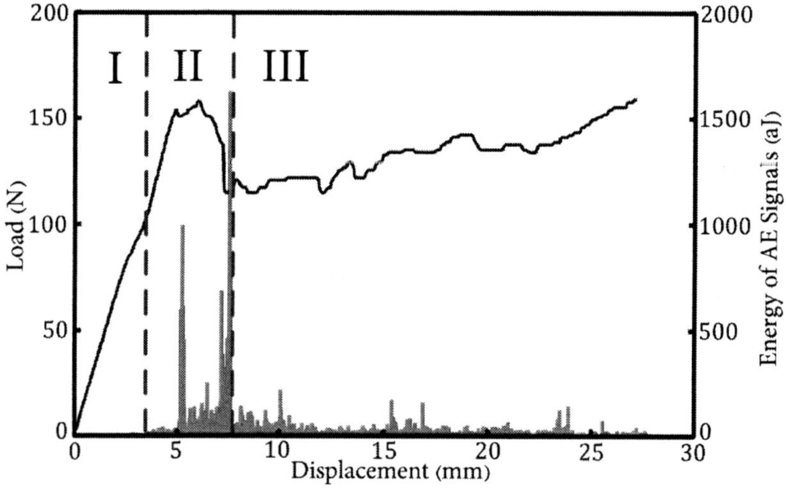

Figure 3. The energy of AE signals, during the initiation and propagation of the delamination, for GW-S1 using MMB setup.

3.1.1. Using Basic Acoustical Features Such as AE Count, AE Energy, AE Cumulative Count, and AE Cumulative Energy for Detecting the Initiation Level

Similar to the mechanical behavior of the specimens, three regions in the load-displacement curves are observable in their AE behavior. For example, these regions are demonstrated for a typical GW-S1specimen in Figure 3. In region I, no substantial AE activity is observed. In region II, by initiation of delamination, the AE activities begin and then increase quickly. In region III, by activation of damage mechanisms and propagation of delamination, the AE activities with a medium level of energy are detected.

The distribution of AE energy and count varies during the crack propagation process. The acoustic emission energy and count distribution throughout the loading process can be used as an indicator method to predict the initiation stage of delamination. This is due to the fact that the initiation stage is a sudden phenomenon releasing high energy that corresponds to the first high rise in energy and count diagrams. The initiation stage of delamination is directly related to the P_C, the load which causes the initiation of delamination. Therefore, for identification of the P_C two methods are applied: a) energy (or count) of the AE signals and b) cumulative energy (or count) of the AE signals, which is the sum of energy or count of the recorded AE signals. In the first approach, the load at the point at which first visible high rise in the energy or count of the AE signals is observed is equal to the P_C.

Figure 4-(a) depicts the P_C acquired by the AE energy method for a typical GW-S1 specimen. In the latter technique, the load at the point at which the first surge emerges in the cumulative energy of the AE signal is equal to the P_C. Figure 4-(b) illustrates the P_C obtained by this approach for a typical GU-S1 specimen.

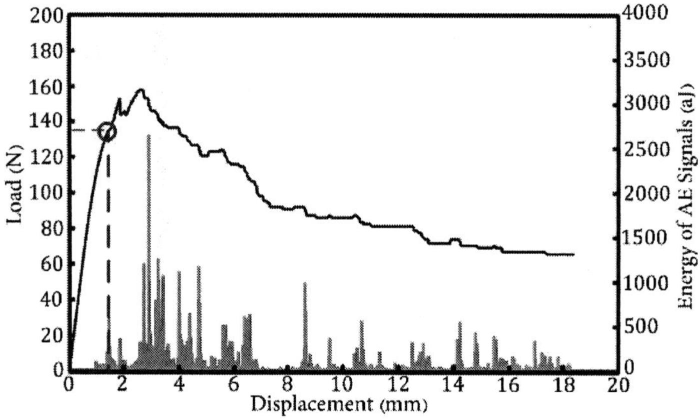

a

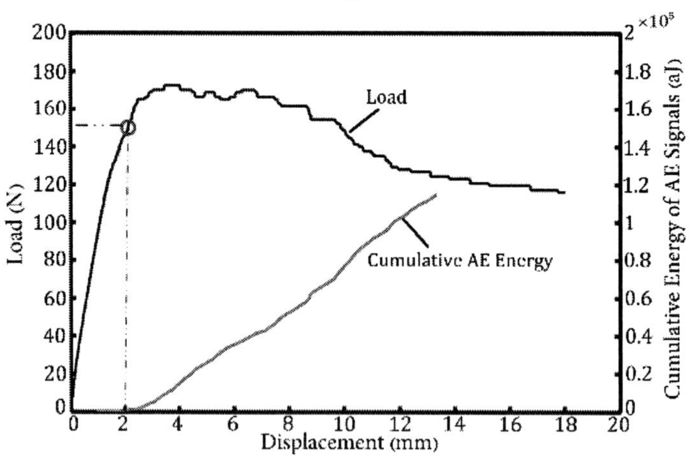

B

Figure 4. (Continued)

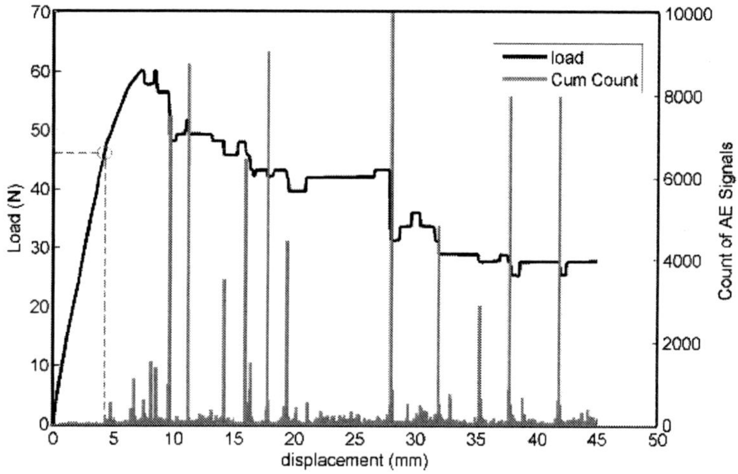

c

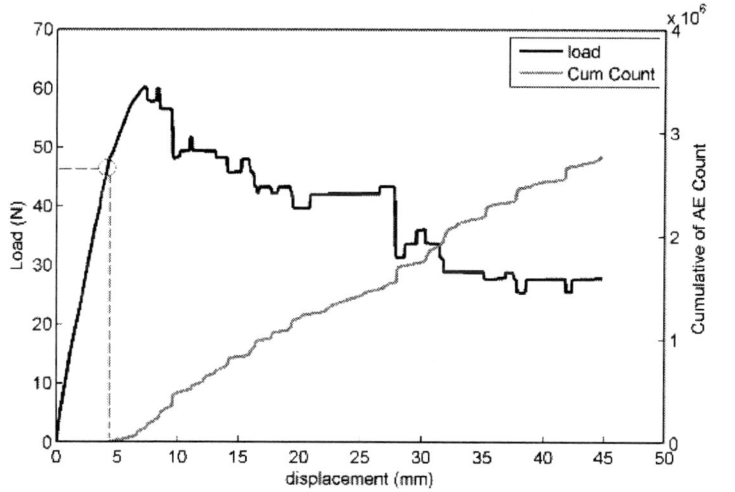

d

Figure 4. Determination of the critical load value using a) AE energy (in GW-S1) and b) the cumulative AE energy (in GU-S1) c) AE count (in GU-S3) and d) cumulative AE count (in GU-S3) approaches using DCB configuration.

Figure 4-(c) shows the distribution of the AE counts during loading for a typical GU-S3 specimen. As can be seen, before the mentioned point, no significant damage is observed and the AE counts are negligible. As the delamination initiates, the load drops off, and the AE counts surge suddenly. When this strain energy reaches to the critical strain energy that is needed for initiation of delamination, the delamination begins and the numbers of the counts grow instantly. The few weak rises in the AE count exist before the initiation of delamination that might be associated with the rubbing and elastic deformation in the specimen. In Figure 4-(d), by considering the cumulative count of the AE signals, at the initiation stage, the delamination starts to propagate as a visible high rise emerges in the cumulative counts of the AE signals. The energy needed for this stage is called critical strain energy. Before this stage, the plot is linear and all the deformations are elastic.

3.1.2. Using Sentry Function That Is a Function Relating the Mechanical and Acoustical Energies, to Obtain the Initiation Stage

In the previous section, AE information and mechanical data were applied separately for the damage characterization. The combination of mechanical data and AE information can also be implemented to have inclusive damage characterization in composite specimens. The function used for this combination is named Sentry function. As indicated by Equation 2, Sentry function is defined as the logarithm form of the ratio of mechanical energy to acoustical energy [12]:

$$f(x) = Ln\left[\frac{E_S(x)}{E_a(x)}\right] \tag{2}$$

Where $E_S(x)$, $E_a(x)$ and x are the strain energy (mechanical energy), the AE events energy and the displacement, respectively. For computation of these energies, two reference volumes were chosen: the volume of the material where the strain energy is stored (V_1) and the volume where the delamination can propagate (V_2) and from which the AE events can take place. Thus, for the Sentry function computation, the strain energy was normalized over the V_1 + V_2 volume and the cumulative acoustic energy was normalized over the V_2 volume. Figure 5 displays the reference volumes in a DCB specimen.

Figure 6 depicts evaluation of the P_C using Sentry function method for a sample specimen. By applying the load, there is a rising trend in the stored strain energy in the specimen. The cumulative AE energy is also low before initiation of delamination (free failure domain), hence, Sentry function has an increasing trend. As the applied load increases, the ability of the material to store strain energy reaches its limits, and the AE cumulative energy significantly increases due to micro-failures appearance, so the slope of the Sentry function trend decreases. During the delamination progression, major failures take place in the specimen and there is an instantaneous release of stored energy due to the internal material failure. This key failure is defined as a significant damage level in delamination and it is in accordance with the initiation of delamination. This is shown by the sudden drop in Sentry function. The load at the point at which the Sentry function trend diminishes instantaneously is equal to the P_C.

For investigation of delamination initiation, the integral of Sentry function can also be used, as defined by Equation 2, where ΩAE is the interval at which the AE signals appear.

$$Int(f) = \int_{\Omega AE} f(x)dx \tag{3}$$

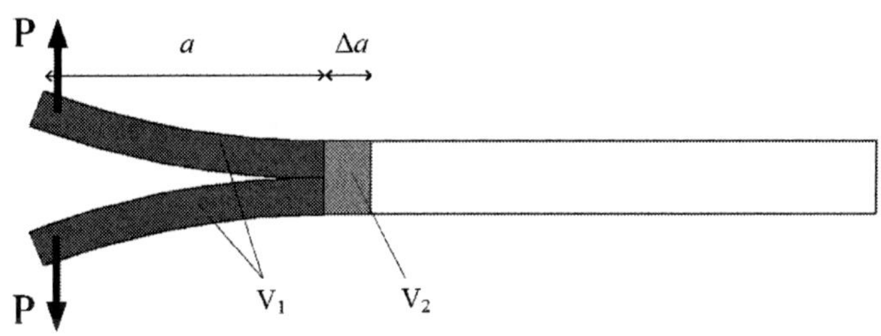

Figure 5. Schematic of the reference volumes in a DCB specimen.

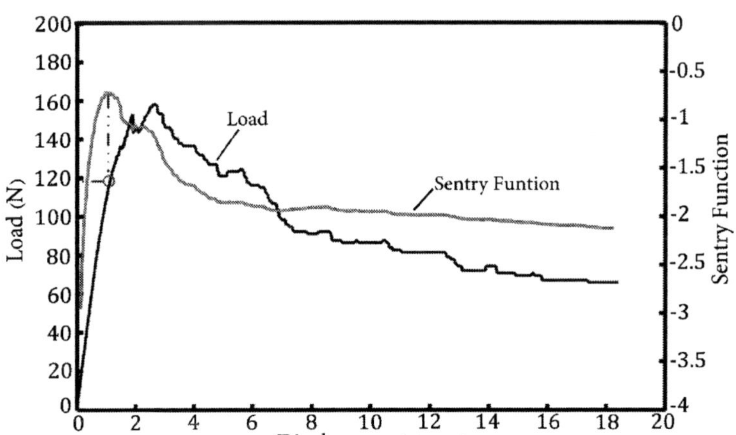

Figure 6. Determination of the critical load value using Sentry function approach for GW-S1 using DCB configuration.

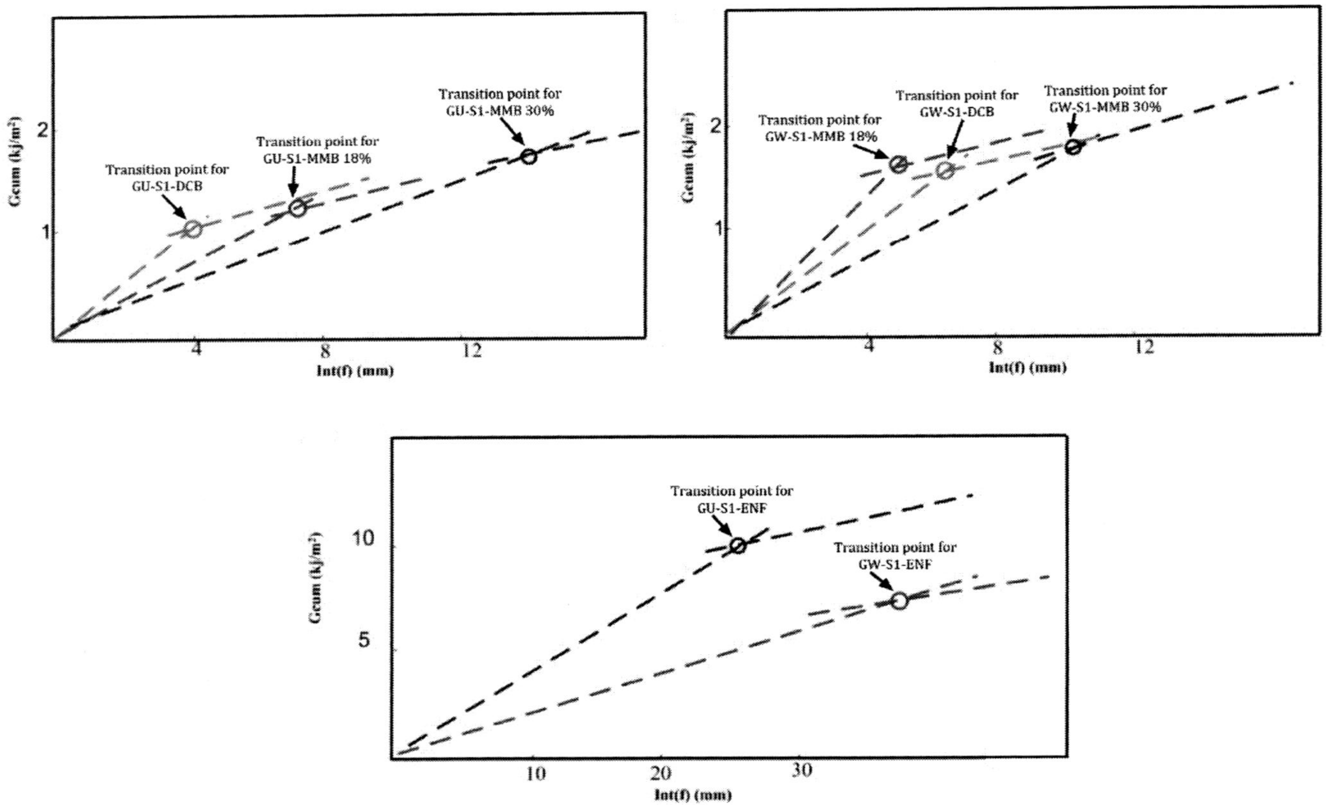

Figure 7. Plot of cumulative G versus integration of Sentry function for the specified specimens.

A distinguished bi-linear relationship between the Int(f) and G_{CUM}, by means of a linear fitting of the data, is applied to evaluate critical strain energy value (as illustrated in Figure 7). G_{CUM} is the sum of the G values computed from the obtained data during the test. Int(f) was also used as a material damage indicator and predictor for the residual strength of the composite laminates. The bi-linear relationship in Figure 7, the appearance of the transition point, is due to a variation in the material strength behavior by altering the delamination crack length. After the transition point, the slope reduces due to decreased crack propagation strength. This behavior implies that when the G_{CUM} is increased by a small amount it makes a significant increase in the Int(f). The slope of the linear diagrams before the transition point is always higher than the slope after the transition point. The slope variation is because of a change of the material strength behavior regarding the onset of delamination. Therefore, the use of the plots in Figure 7 allows evaluating the G_C values.

3.2. Progression of Delamination

Determination of the amount of delamination crack growth has a vital role in the damage tolerance analyses of the composite structures [18, 19]. In order to predict delamination growth in laminated composites, there are two main approaches; one is based on the localization of the AE signals, which is originated from the damaged area of the crack tip, and the other one is based on the AE energy released due to the delamination crack propagation.

3.2.1. Delamination Growth Prediction Using AE Signals Localization

According to the literature review [19-21], most of the damage mechanisms in laminated composites such as fiber breakage, matrix cracking, and fiber/matrix debonding occur at the small region at the crack tip, where stress is concentrated. Thus, if the velocity of AE wave in the composite material is specified, the coordinates of the crack tip can be determined using a simple AE localization method. To this aim, first, the AE wave velocity in the composite material should be specified. The velocity of AE wave in the materials is calculated according to the standard pencil lead breakage test method ASTM E976-10 [22]. The AE wave velocity in different composite materials is represented in Table 2.

Table 2. The AE wave velocity in different composite materials

Specimen	AE wave velocity (m/s)
Unidirectional Carbon/Epoxy	4937 ± 3.0%
Unidirectional Glass/Epoxy	4668 ± 1.0%
Woven (0/90) Glass/Epoxy	3380 ± 0.5%

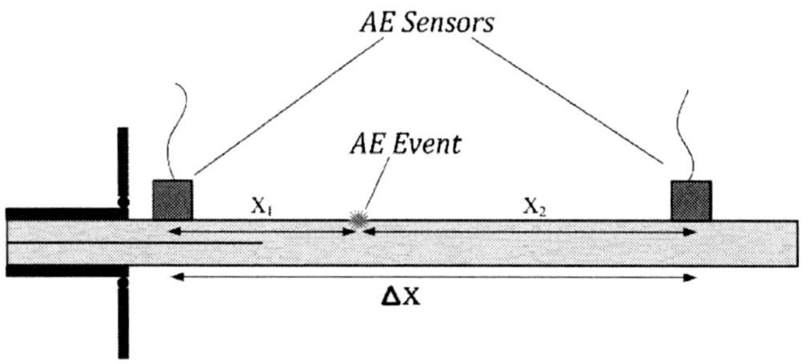

Figure 8. The schematic of AE signals localization [8].

Linear localization is utilized to predict delamination growth. In this method, two AE sensors are placed on two ends of the specimens, so that, the initial crack tip is placed between the two sensors (Figure 8). If a damage occurs in the space between the sensors, the originated AE wave reaches to the sensors at different times. This is due to the different distances of the damage from the sensors. By determining the time difference (Δt), distance between the AE sensors (Δx), and AE wave velocity (V) in the laminate, the location of the damage can be specified using Equation 3:

$$\begin{cases} x_1 + x_2 = \Delta x \\ |x_2 - x_1| = V.(t_2 - t_1) \end{cases} \quad ; \quad \Delta t < \frac{\Delta x}{V} \tag{3}$$

The delamination growth results using the localization approach are represented in Figure 9.

As can be seen, AE could predict the general trend of delamination growth, but the scattering of the AE results is too high. To reduce the scattering and increase the accuracy of the AE prediction, a new methodology was proposed. It has been shown that, in most of the laminated composites, fiber-bridging phenomena occurs at the small region behind the crack tip (see Figure 10) [23-26].

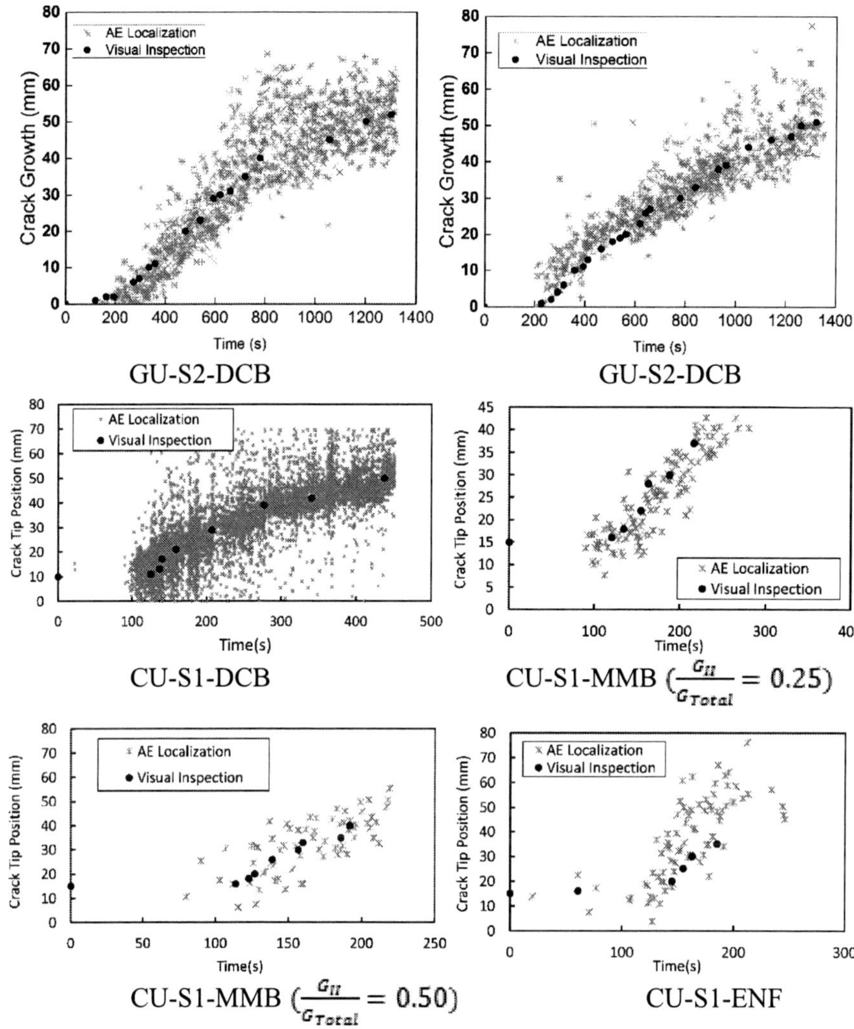

Figure 9. Prediction of delamination growth using the AE localization method.

Therefore, if the AE signals originated from the fiber breakage could be separated from the other AE signals, and the localization is being carried out using these signals, the accuracy of the AE prediction will be increased. To this aim, the fiber bundle tensile test was carried out and the originated AE signals were recorded (Figure 11). Then, the frequency range of the fiber breakage signals was specified using Fast Fourier Transform (FFT), [350 kHz-500 kHz] (see Figure 12). According to Figure 13, by eliminating the AE signals with frequency content out of this range, the scattering of the AE data is decreased. As fiber breakage occurs behind the crack tip, the upper bound of the AE data represents the crack tip position.

However, in the specimens that no fiber bridging occurs, the last procedure cannot be used. To increase the accuracy of the AE prediction for these specimens, a new methodology is proposed. As illustrated in Figure 14, in this method, time domain of the test is discriminated to a few sections. In each section, there are a number of AE events with proprietary times and positions. All the AE signals within this section should be replaced with one AE event, which its time and position are the mean values of the times and positions of the AE events that are within the section. The appropriate number of sections is obtained by trial and error. The accuracy of the AE prediction is not changing by changing the number of sections in a great range (from 15 to 100 sections).

3.2.2. Prediction of Delamination Crack Growth Using Cumulative AE Energy

Figure 15 shows the delamination crack growth and cumulative AE energy of the unidirectional glass/epoxy DCB specimen (GU-S1).

As can be seen, the cumulative AE energy and delamination crack growth curves have similar trends. Thus, a linear relation between the cumulative AE energy and delamination crack growth can be established (Figure 16). Using this linear relation, the history of the delamination crack growth can be predicted using the recorded AE signals as represented in Figure 17.

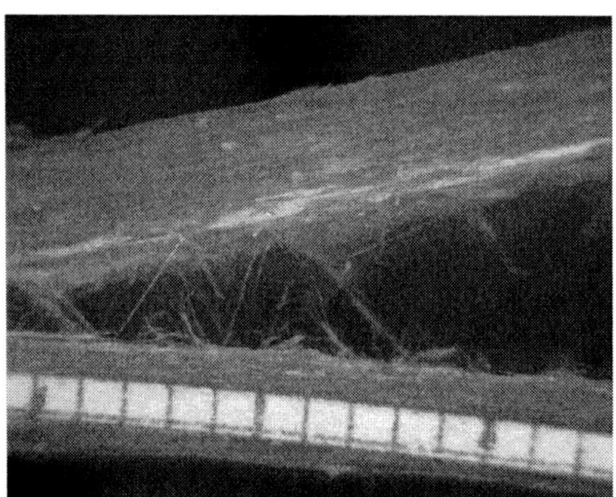

Figure 10. The fiber bridging phenomenon for GU-S1.

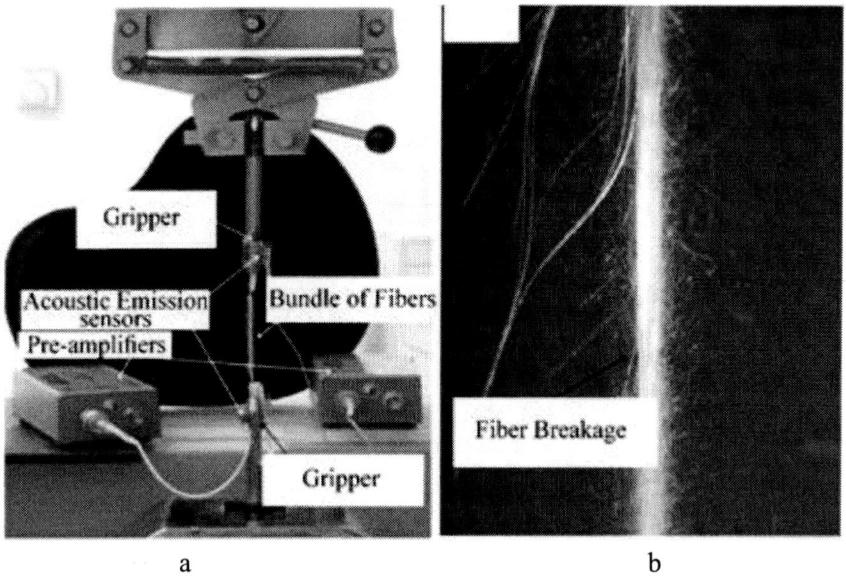

Figure 11. The glass fiber bundle tensile test.

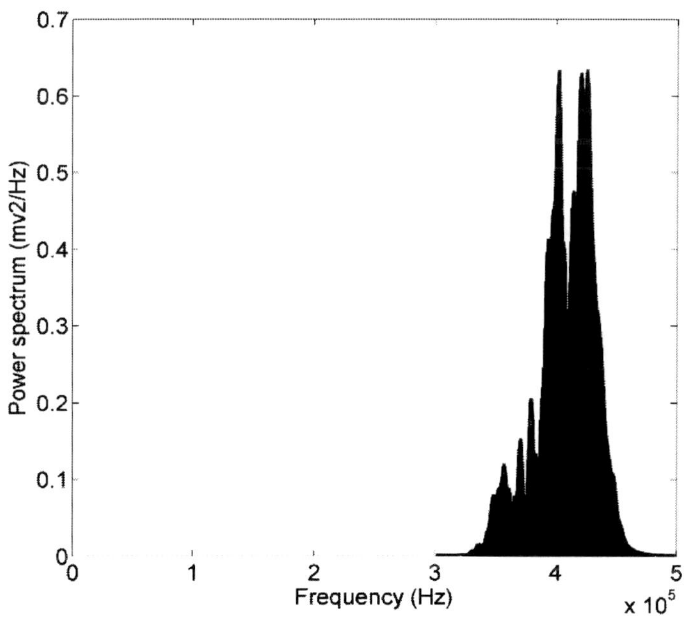

Figure 12. FFT results showing the frequency range of the AE signals obtained from the fiber bundle tensile test.

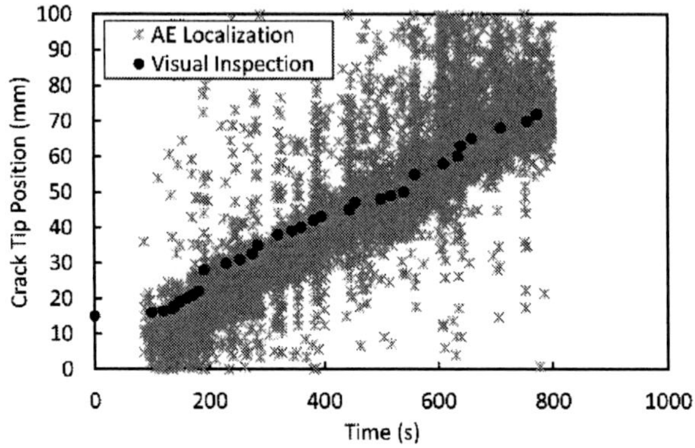

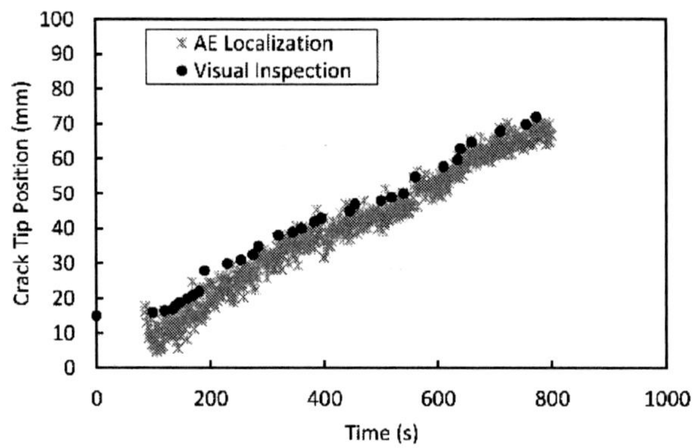

Figure 13. Increasing the accuracy of delamination growth using the AE method.

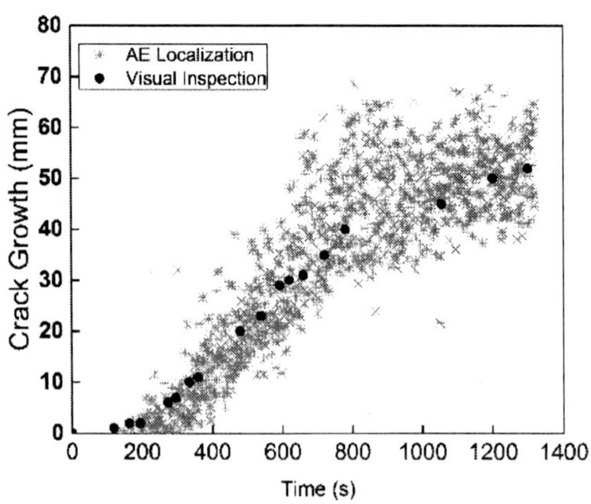

Primary prediction of AE for GU-S2-DCB

Figure 14. (Continued)

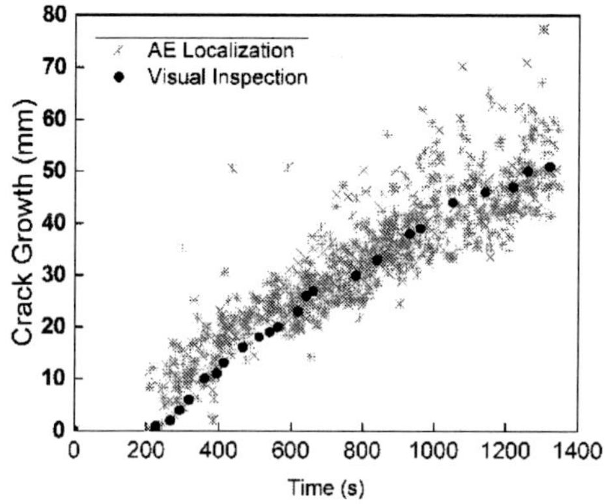

Primary prediction of AE for GW-S2-DCB

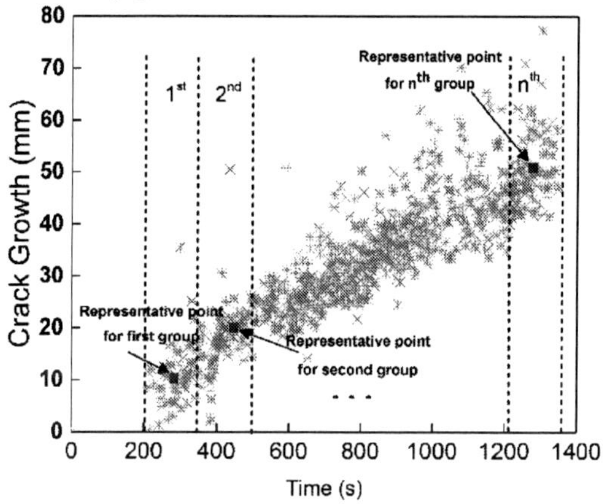

Discrete of the time domain of the test to a number of sections

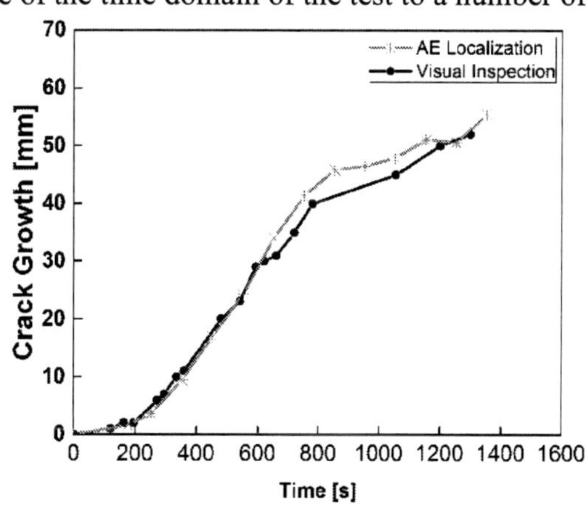

Final prediction of AE for GU-S2-DCB

Figure 14. (Continued)

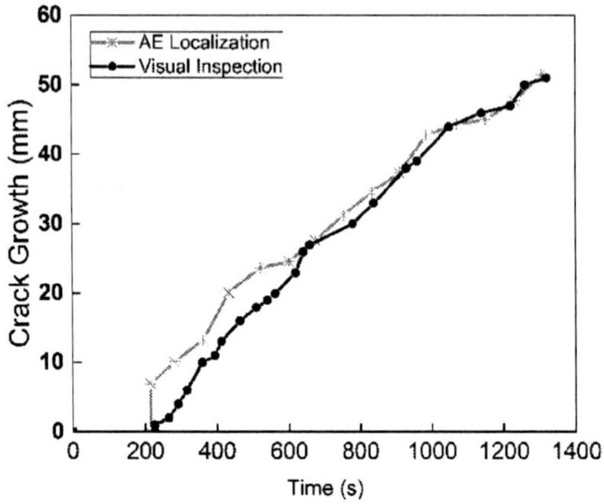

Final prediction of AE GW-S2-DCB

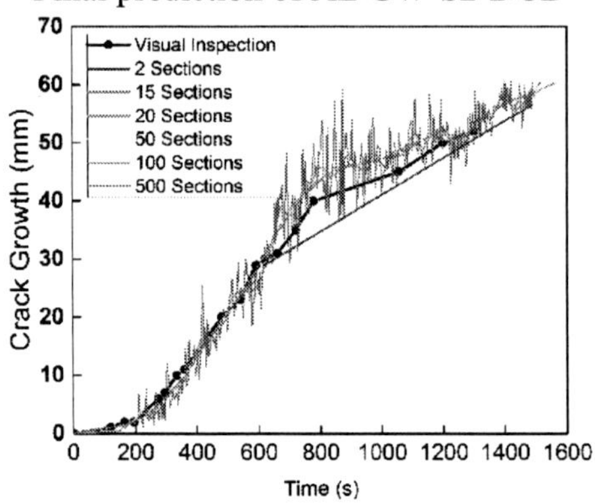

The effect of the number of sections on the predicted results for GU-S2-DCB

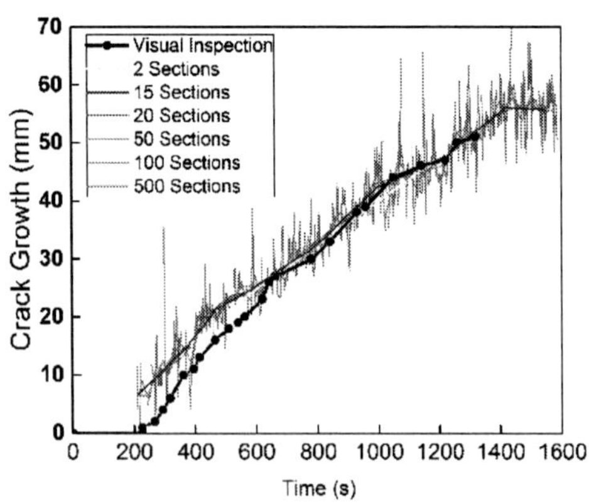

The effect of the number of sections on the predicted results for GW-S2-DCB

Figure 14. Prediction of delamination growth using AE method in the specimens without any fiber bridging.

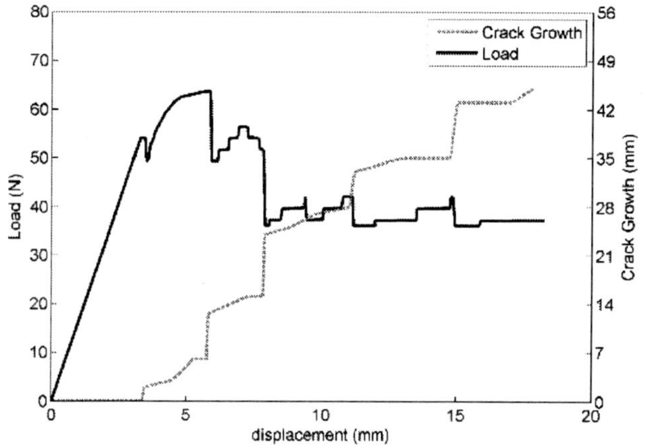

Crack growth curve (GU-S3-DCB)

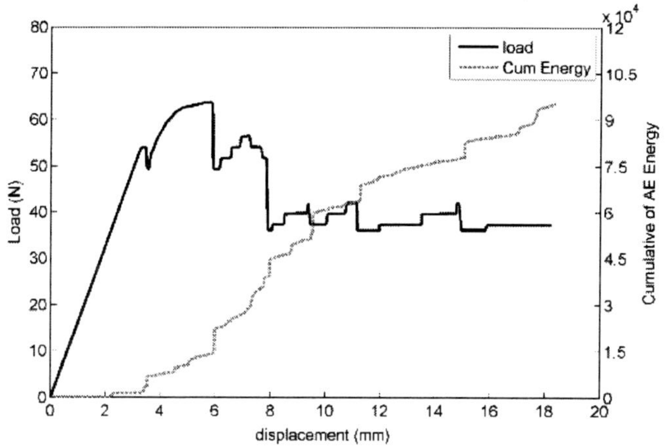

Cumulative AE energy curve

Figure 15. Crack growth and cumulative AE energy of the unidirectional glass/ epoxy specimens.

In order to insure the applicability of the method under various conditions, it is also applied to measure the crack length for the carbon/epoxy composites under different loading conditions (Figure 18). The obtained results showed a very interesting consistency with the visually measured crack growth length; this is especially very useful in the case of mode II loading condition where the crack growth is unstable and hard to measure.

This method could predict delamination propagation very well, but in the early stage of the delamination growth, the method overestimates the delamination length. The reason is that, at the beginning of the crack growth, the matrix cracking, that has a lower AE energy content, is the dominant damage mechanism. Thus, in this stage, the relation of the cumulative AE energy and crack growth is not linear (Figure 19). By progression of the delamination, other damage mechanisms such as fiber breakage and fiber/ matrix debonding are activated, which have higher AE energy. Consequently, the relation between cumulative AE energy and the crack growth is close to the linear relationship. By curve fitting, it is clear that there is a power relationship between the cumulative AE energy and delamination crack growth. The proposed model could predict all stages of the delamination growth very well (Figure 20).

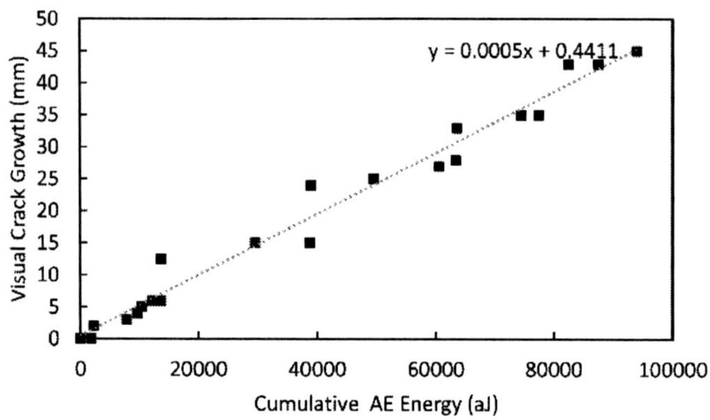

Figure 16. The linear relationship between the delamination crack growth and cumulative AE energy for GU-S3-DCB.

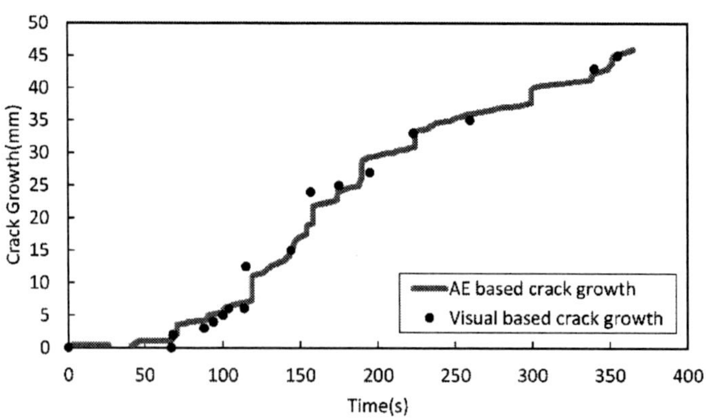

Figure 17. Prediction of the delamination crack growth using the cumulative AE energy for GU-S3-DCB.

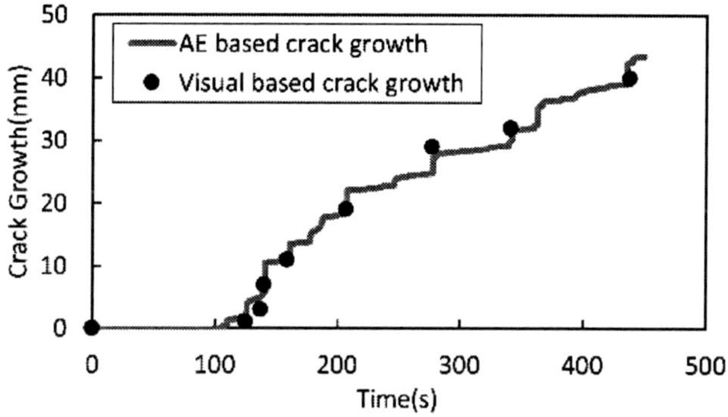

CU-S1-DCB

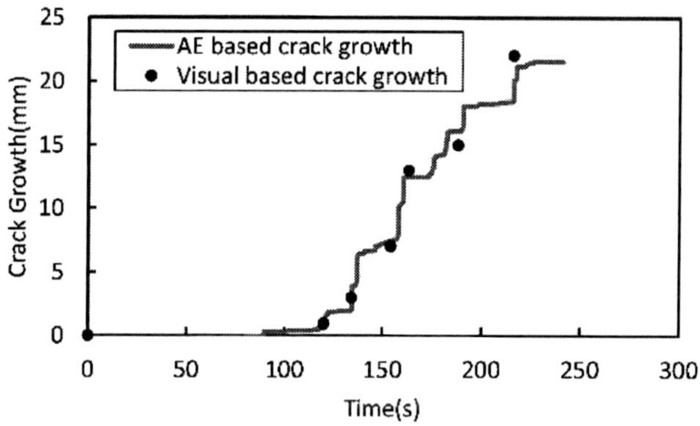

CU-S1-MMB ($\frac{G_{II}}{G_{Total}} = 0.25$)

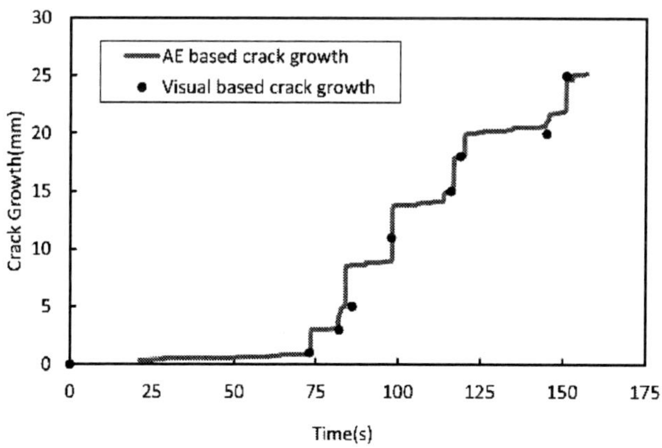

CU-S1-MMB ($\frac{G_{II}}{G_{Total}} = 0.50$)

Figure 18. (Continued)

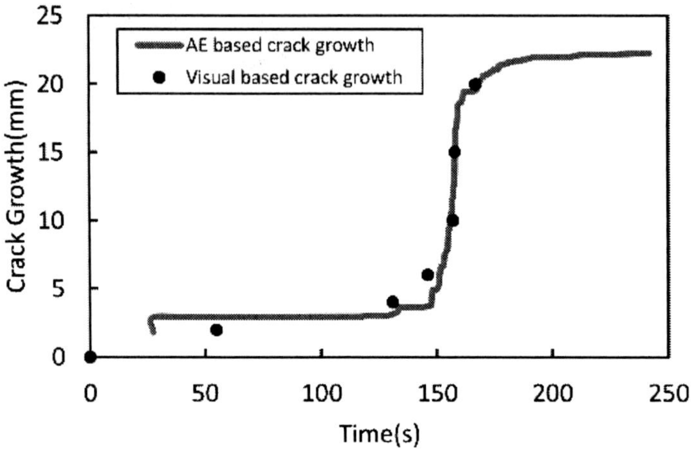

CU-S1-ENF

Figure 18. Prediction of delamination crack growth using cumulative AE energy in unidirectional carbon/epoxy specimens, under mode I, mode II and mixed-mode I&II loading conditions.

3.2.3. Damage Clustering

Wavelet Packet Transform (WPT) and Fuzzy C-Means (FCM) clustering methods are used to cluster the damage mechanisms that occurred in the specimens during delamination propagation. Prior to use of these methods, first, the AE features of each damage mechanism must be specified. To this aim, the fiber bundle and pure resin tensile tests were carried out and the AE signals of these damage mechanisms are recorded (Figure 20 and Figure 21). Then, using FFT analysis, the frequency range of each damage mechanism is specified and are utilized as AE reference patterns in the clustering procedure (Figure 12 and Figure 22). The flowcharts of the clustering procedures are presented in Figure 23.

3.2.3.1. Wavelet Packet Transform

Wavelet Transform (WT) is one of the advanced methods for signal processing. Some studies during the last decade utilized the WT as an appropriate technique for analyzing AE signals [27-29]. Fundamental and applications of WT were noted in the previous studies [30-31]. Discrete Wavelet Transform (DWT) is utilized to decompose the AE signals into some components (approximation and detail components). Mathematically, DWT is defined as:

$$f(t) = c\sum_{i}\sum_{k} DWT(i,k)2^{\frac{-i}{2}}\psi(2^{-i}t - k) \tag{6}$$

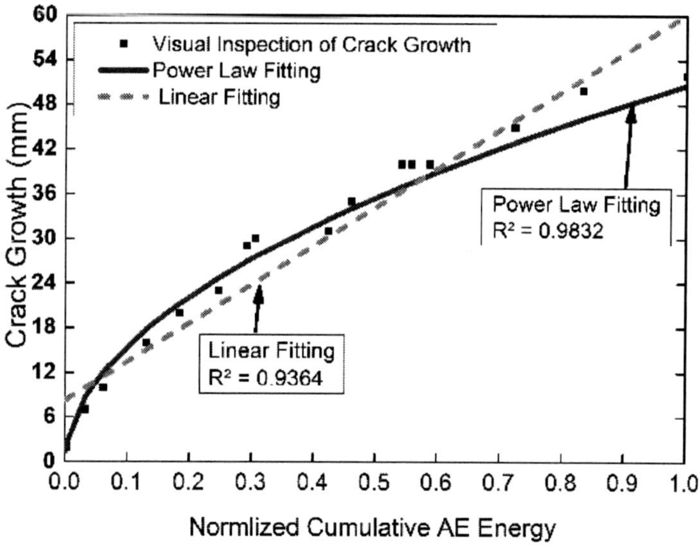

GU-S2-DCB

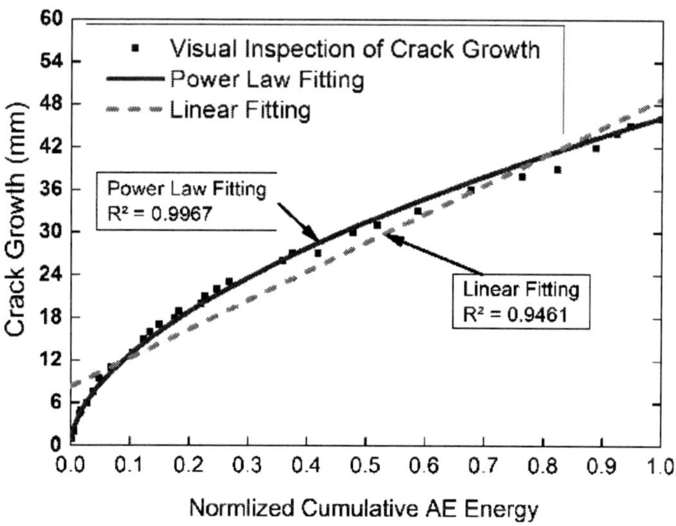

GW-S2-DCB

Figure 19. The relation between cumulative AE energy and delamination crack growth.

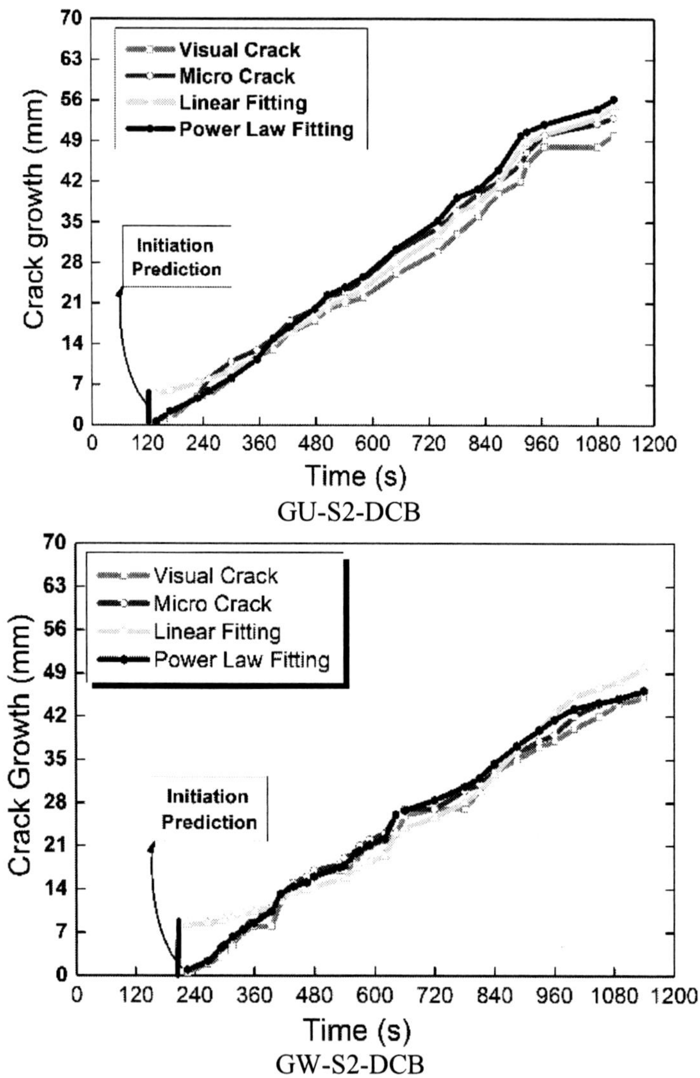

Figure 20. Prediction of delamination crack growth using linear and power relation between the cumulative AE energy and delamination crack growth.

The inverse of DWT is expressed as:

$$DWT(i,k) = \int_{-\infty}^{+\infty} f(t) 2^{\frac{i}{2}} \psi^* (2^i t - k)\, \mathrm{d}t \tag{7}$$

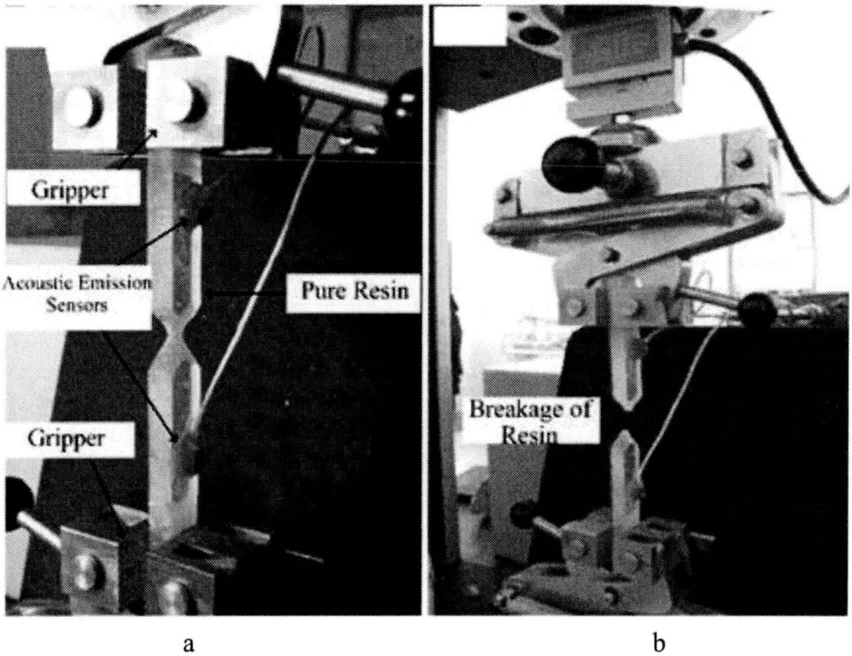

Figure 21. Tensile test of the pure resin.

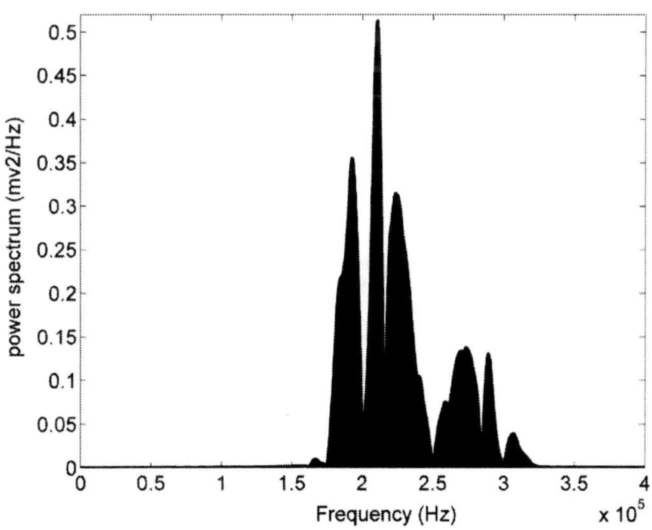

Figure 22. The frequency range of the matrix cracking obtained from FFT.

In the above relations, f(t), i, DWT (i,k), k, and ψ, are the signal that should be analyzed the decomposition level, the wavelet transform coefficients, the time domain and a mother wavelet, respectively.

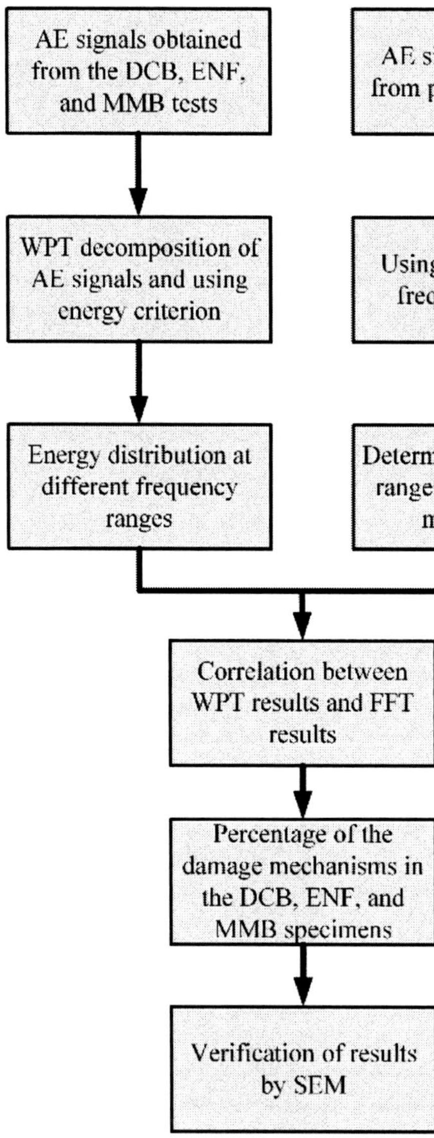

Clustering with WPT

Figure 23. (Continued)

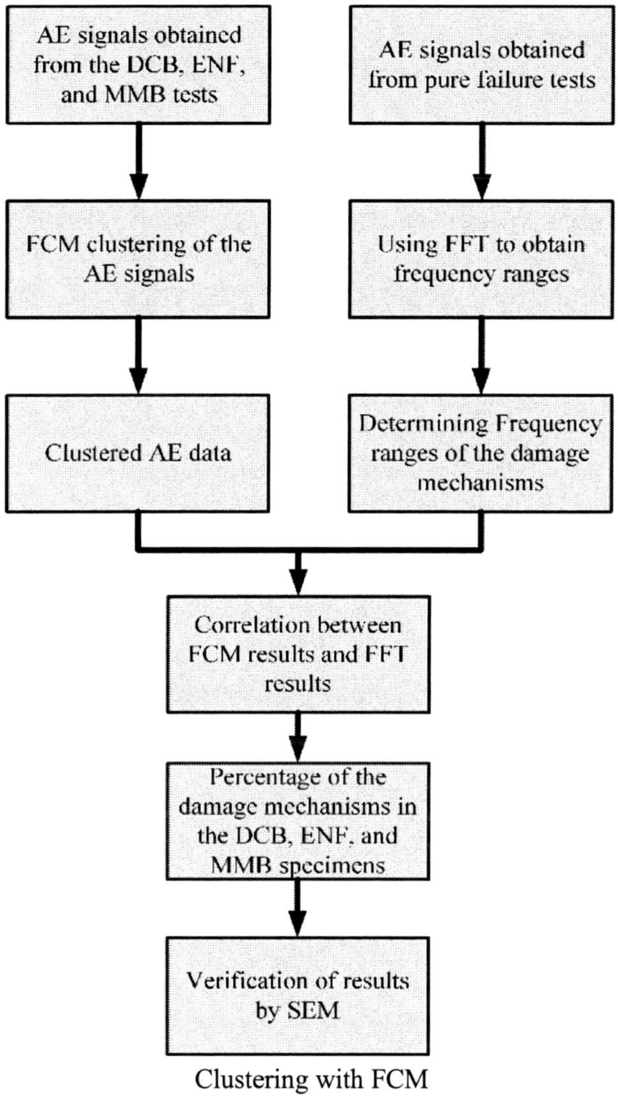

Figure 23. The flowchart of the damage clustering.

A drawback of DWT is that this method neglects the high frequency components of the AE signals during the decomposition process. Thus, some of the valuable information is eliminated [32]. To overcome this problem, WPT is utilized.

In this method, both components of the decomposed signal (the components and details) are decomposed in the decomposition process.

Wavelet packet transform is defined by $\psi^{i}_{j,k}$ as follow:

$$\psi^{i}_{j,k}(t) = 2^{-j/2}\psi^{i}(2^{-j}t - k) \tag{8}$$

Where i, j, and k are modulation, scale, and translation parameters, respectively.

The wavelet function ψ^{i} is obtained from the following equations:

$$\psi^{2i}(t) = \frac{1}{\sqrt{2}}\sum_{k=-\infty}^{\infty} h(k)\psi^{i}(\frac{t}{2} - k) \tag{9}$$

$$\psi^{2i+1}(t) = \frac{1}{\sqrt{2}}\sum_{k=-\infty}^{\infty} g(k)\psi^{i}(\frac{t}{2} - k) \tag{10}$$

Where h(k) and g(k) are the quadrature mirror filters accompanying with the scaling function and the mother wavelet function.

The wavelet packet coefficients $c^{i}_{j,k}$ are defined by:

$$c^{i}_{j,k} = \int_{-\infty}^{+\infty} f(t)\psi^{i}_{j,k}(t)\,dt \tag{11}$$

The wavelet packet component $f^{i}_{j}(t)$ is calculated as follows:

$$f^{i}_{j}(t) = \sum_{k=-\infty}^{+\infty} c^{i}_{j,k}\psi^{i}_{j,k}(t)\Delta t \tag{12}$$

The frequency range of each wavelet packet component is defined as follows.

The frequency range of the component:

$$[0, \frac{1}{2}f_{s}2^{-j}] \tag{13}$$

The frequency range of the detail:

$$[\frac{1}{2}f_s 2^{-j}, \frac{1}{2}f_s 2^{-(j-1)}] \tag{14}$$

The energy of each component is calculated by:

$$E_j^i(t) = \sum_{\tau=t_0}^{t} \left(f_j^i(\tau)\right)^2 \tag{15}$$

In this study, to determine the energy content of each component, the ratio of the energy of each component to total energy is calculated using:

$$P_j^i(t) = \frac{E_j^i(t)}{E_{Total}(t)} \qquad i = 1,\ldots,2^j \tag{16}$$

Where $P_j^i(t)$ is the relative distributed energy.

The results of WPT for the damage clustering in the unidirectional and woven glass/epoxy DCB, ENF and MMB specimens are represented in Figure 24 and Table 3.

Table 3. Damage clustering of the DCB, ENF, and MMB glass/epoxy specimens using WPT

		Frequency ranges (kHz)		
	Specimen	187.5-250 (Matrix cracking)	250-312.5 (Fiber/matrix debonding)	375-437.5 (Fiber breakage)
Percentage of energy	GU-S1-DCB	25	20	26
	GU-S1- MMB 18%	30	26	19
	GU-S1- MMB 30%	40	18	18
	GU-S1- ENF	47	26	7
	GW-S1- DCB	35	29	8
	GW-S1-MMB 18%	36	31	8
	GW-S1-MMB 30%	38	32	7
	GW-S1-ENF	52	23	7

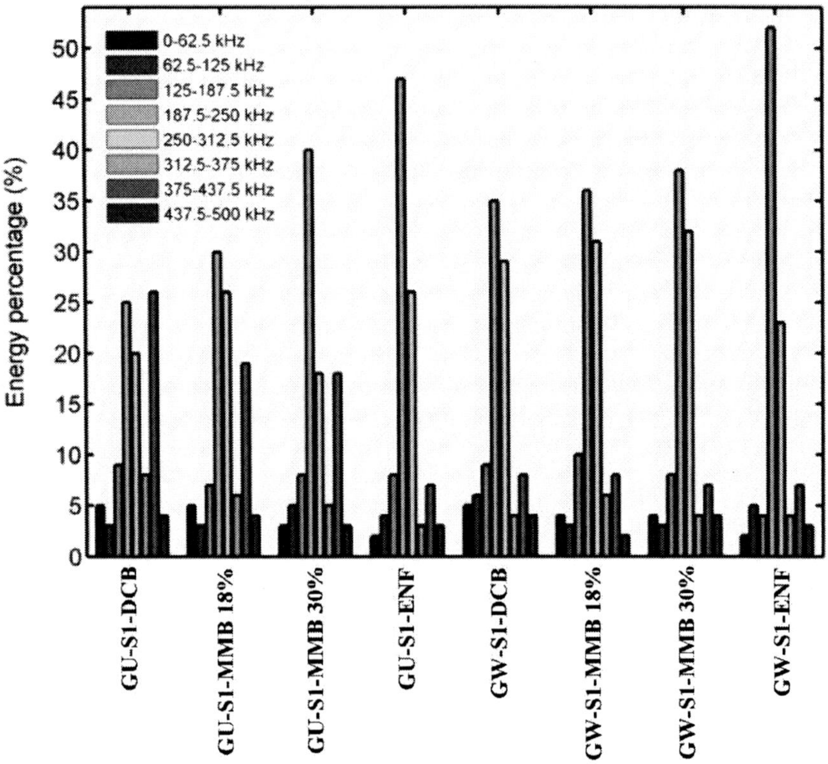

Figure 24. Damage clustering of the DCB, ENF, and MMB glass/epoxy specimens using WPT.

3.2.3.2. Fuzzy C-Means Clustering

Classifying data into groups whose members are similar in some way is defined as clustering process. Clustering methodologies are categorized into two types, including hard clustering and fuzzy clustering approaches. Hard clustering techniques comprise classical set theory and need that an object either does or does not belong to a specific cluster, but fuzzy clustering techniques permit the objects to belong to some other clusters simultaneously, with different degrees of membership between 0 and 1 [33, 34].

The data are the observations of some physical phenomena. Each observation consists of n measured variables, separated into an n-dimensional column vector $p_r = [p_{1r}, ..., p_{nr}]^T$. A set of m observations is denoted by $P = \{p_r | r = 1,2,...,m\}$, and is expressed as an n × m matrix [33, 34]:

$$P_{n\times m} = \begin{bmatrix} p_{11} & \cdots & p_{1m} \\ \vdots & \ddots & \vdots \\ p_{n1} & \cdots & p_{nm} \end{bmatrix} \tag{3}$$

According to membership functions, a partition can be easily presented by the partition matrix $U = [\gamma_{ir}]_{c\times m}$. The components of U must satisfy the subsequent conditions:

$$\gamma_{ir} \in [0,1] \quad 1 \le i \le c, 1 \le r \le m \tag{4}$$

$$\sum_{i=1}^{c} \gamma_{ir} = 1 \quad 1 \le r \le m \tag{5}$$

$$0 \prec \sum_{r=1}^{m} \gamma_{ir} \prec m, \quad 1 \le i \le c \tag{6}$$

it can be revealed that partition matrix (U) has c rows and m columns; c is the number of clusters, m is the number of observations. The data matrix has n rows and m columns and n is the dimension of each observation where c < n. Each element of U has a value between 0 and 1. The sum of elements on each column of U is 1.

For clustering the AE data, FCM with three clusters was applied. The number of clusters in each data set was determined due to the fact that generally within the glass/epoxy composite materials three damage mechanisms (i.e., matrix cracking, fiber/matrix debonding and fiber breakage) are common. The outcome of the analysis using three classes with different central characteristic is shown in Figure 25. The FCM results show that the AE signals are well separated along the first principal direction and are clustered in three different classes. Moreover, it can be observed from Table 2 that the AE signals distribution percentage at each cluster varies from one experimental condition to another. By considering the AE parameters of the obtained classes, the frequency parameter was best distinguished. These figures highlight three separate frequency ranges for these specimens. The lower band frequency class (class1) is the earliest damage mode that appears in the tests.

The obtained results from WPT and FCM are compared with SEM images of the damaged surfaces of the specimens and good consistency is observed between the WPT results and SEM images (Figure 26).

Table 2. Damage clustering of DCB, ENF, and MMB glass/epoxy specimens using FCM

Specimen	Percentage of each damage mechanism		
	First class (Matrix cracking)	Second class (Fiber/matrix debonding)	Third class (Fiber breakage)
GU-S1-DCB	34	30	36
GU-S1-MMB 18%	40	35	25
GU-S1-MMB 30%	55	23	22
GU-S1-ENF	60	30	10
GW-S1-DCB	49	38	13
GW-S1-MMB 18%	51	37	12
GW-S1-MMB 30%	53	37	10
GW-S1-ENF	65	27	8

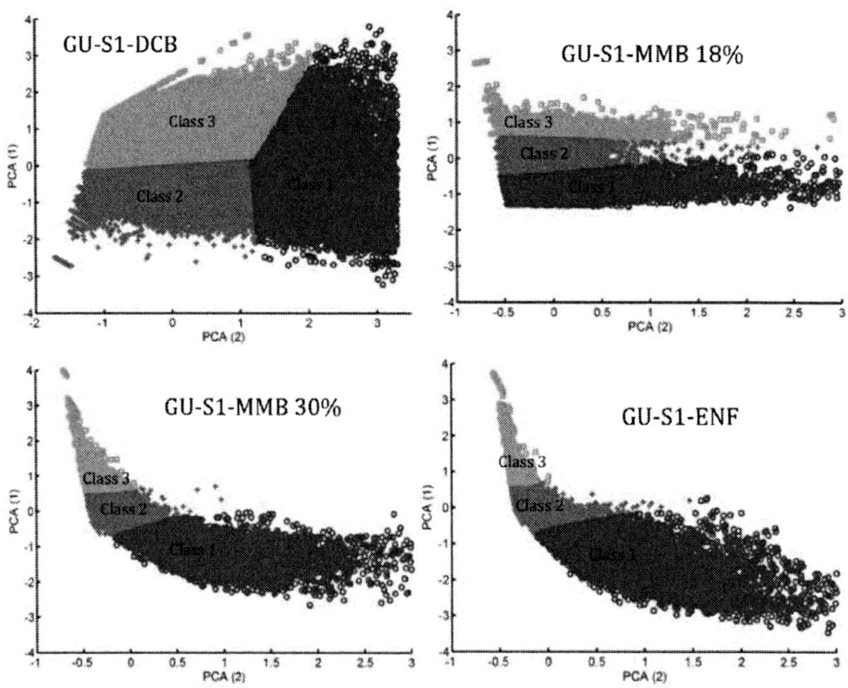

Figure 25. (Continued)

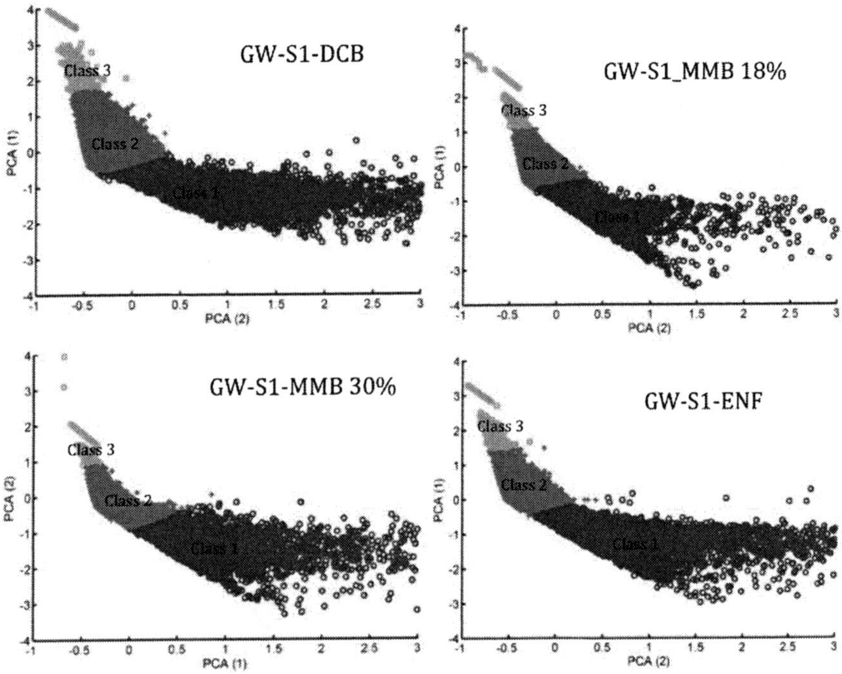

Figure 25. Damage Clustering using FCM approach.

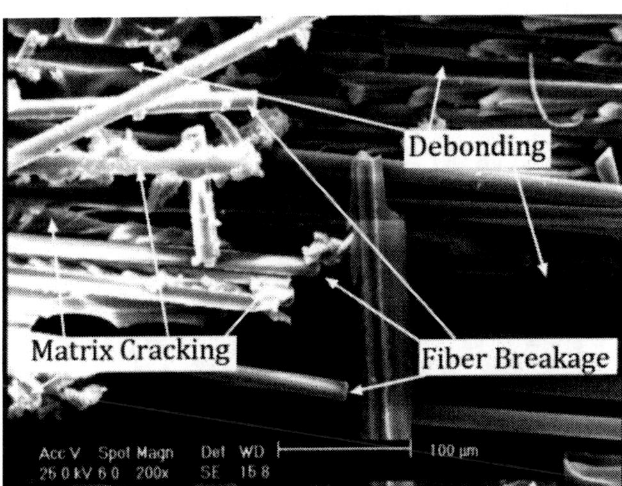

GU-S1-DCB

Figure 26. (Continued)

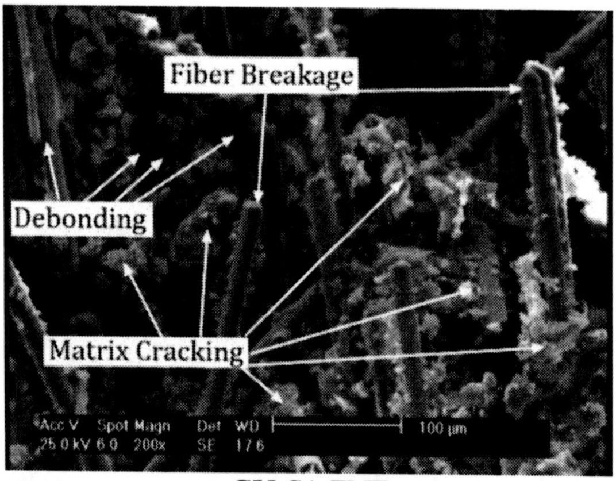

GU-S1-ENF

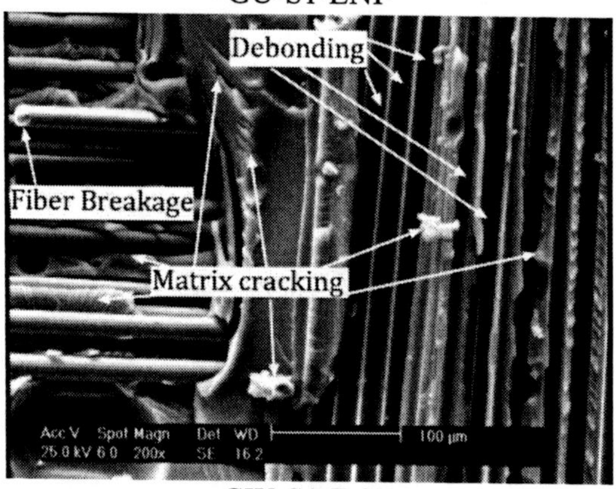

GW-S1-DCB

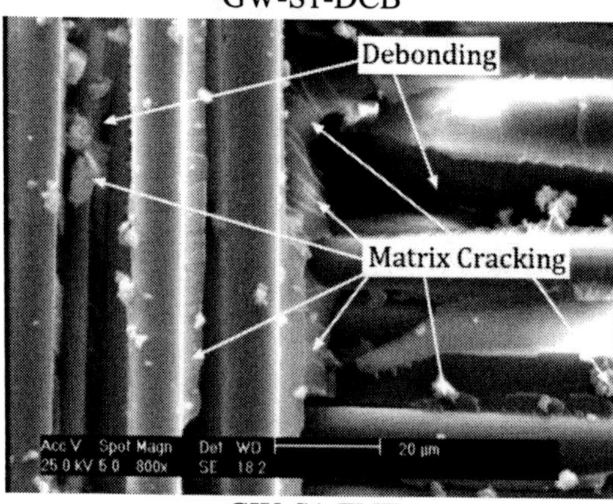

GW-S1-ENF

Figure 26. SEM images of damaged surfaces of unidirectional and woven glass/epoxy composites under different loading conditions.

CONCLUSION

This work summarizes the potential of AE technique to characterize the initiation and propagation of delamination in laminated composites. Carbon/ epoxy and glass/epoxy specimens subjected to mode I (DCB), mode II (ENF), and mixed-mode I and II (MMB) loading conditions are studied. The following items are concluded.

1) The obtained results from the introduced AE methods to detect initiation of delamination, especially the Sentry function method, are shown to be more conservative than those derived from conventional fracture mechanics methodologies. This is due to the fact that AE monitoring is able to detect micro-damage mechanisms that occur before the macroscopic delamination initiation.
2) A very good agreement is achieved between the AE-predicted crack length and the actual crack length observed visually.
3) Scanning electron microscopic observations from the damaged mechanisms are in a good agreement with the AE-based classification results obtained by WPT and FCM methods. Therefore, it is concluded that the presented methods are successful in the classification process to improve the characterization of the damage mechanisms, in actual occurring modes of delamination, that is, mode I, mode II and the combination of these pure modes.
4) The SEM observations revealed that the damage mechanisms of matrix cracking, debonding, and fiber breakage were the sources of the AE signals.

REFERENCES

[1] Paris, I., and Poursartip, A. 1998. "Delamination crack tip behavior at failure in composite laminates under mode I loading." *Journal of Thermoplastic Composite Materials* 11(1):57-69.

[2] Ho-Cheng, H., and Dharan, CKH. 1990. "Delamination during drilling in composite laminates." *Journal of Engineering for Industry* 112(3):236-239.

[3] Fotouhi, M., and Ahmadi, M. 2014. "Acoustic emission-based study to characterize the initiation of delamination in composite materials." *Journal of Thermoplastic Composite Materials* 29(4):519-537.

[4] Tay, T.E. 2001. "Characterization and analysis of delamination fracture in composites: An overview of developments from 1990 to 2001." *Applied Mechanics Reviews* 56(1):1-32.

[5] Benzeggagh, M.L., and Kenane, M. 1996. "Measurement of mixed-mode delamination fracture toughness of unidirectional glass/epoxy composites with mixed-mode bending apparatus." *Composites Science and Technology* 56(4):439-449.

[6] Barre, S., and Benzeggagh, M.L. 1994. "On the use of acoustic emission to investigate damage mechanisms in glass-fiber reinforced polypropylene." *Composite Science Technology* 52:369-376.

[7] Bakukas, J., Prosser, W., and Johnson, W. 1994. "Monitoring damage growth in titanium matrix composites using Acoustic Emission." *Journal of Composite Materials* 28(4):305-328.

[8] Silversides, I., Maslouhi, A., and LaPlante, G. 2013. "Acoustic emission monitoring of interlaminar delamination onset in carbon fibre composites." *Structural Health Monitoring* 12(2):126-140.

[9] ASTM Standard E610-98A. 1991. "Definitions of terms relating to acoustic emission." West Conshohocken, PA: ASTM International.

[10] Miller, R.K. 1987. "Nondestructive Testing Handbook: Acoustic Emission Testing." 5th ed., American Society for Nondestructive Testing.

[11] ASTM Standard D5528. 2002. "Standard test method for mode I interlaminar fracture toughness of unidirectional fiber-reinforced polymer matrix composites." West Conshohocken, PA: ASTM International.

[12] De Kalbermatten, T., Jaggi, R., Flueler, P. et al. 1992. "Microfocus radiography studies during mode I interlaminar fracture toughness tests on composites." *Journal of material science letters* 11(9):543-546.

[13] Sridharan, S. 2008. "Delamination behaviour of composites." New York: CRC Press.

[14] Tan, P., and Tong, L. 2007. "Experimental and analytical identification of the delamination using isolated PZT sensor and actuator patches." *Journal of Composite Materials* 41(4):477-492.

[15] Benmedakhene, S., Kenane, M., and Benzeggagh, M.L. 1991. Initiation and growth of delamination in glass/epoxy composites subjected to static and dynamic loading by acoustic emission monitoring. *Composite Science Technology* 59:201-208.

[16] Bohse, J., and Brunner, A.J. 2008. "Acoustic emission in delamination investigation. In: Sridharan S (ed.) *Delamination behaviour of composites*. New York: CRC Press, 217-277.

[17] Minak, G., and Zucchelli, A. 2008. "Damage evaluation and residual strength prediction of CFRP laminates by means of acoustic emission techniques." In: Durand LP (ed.) *Composite materials research progress*. New York: Nova Science Publishers Inc, 165-207.

[18] Saeedifar, M., Fotouhi, M., Ahmadi Najafabadi, M., Hosseini Toudeshky, H., and Minak, G. 2016. "Prediction of quasi-static delamination onset and growth in laminated composites by acoustic emission." *Composites Part B* 85:113-122.

[19] Saeedifar, M., Fotouhi, M., Ahmadi Najafabadi, M., Hosseini Toudeshky, H. 2015. "Prediction of delamination growth in laminated composites using acoustic emission and Cohesive Zone Modeling techniques." *Composite Structures* 124:120-127.

[20] Mohammadi, R., Saeedifar, M., Hosseini Toudeshky, H., Ahmadi Najafabadi, M., and Fotouhi, M. 2015. "Prediction of delamination growth in carbon/epoxy composites using a novel acoustic emission-based approach." *Journal of Reinforced Plastics and Composites* 34(11):868-878.

[21] Saeedifar, M., Ahmadi Najafabadi, M., Yousefi, J., Mohammadi, R., Hosseini Toudeshky, H., Minak, G. 2017. "Delamination analysis in composite laminates by means of Acoustic Emission and bi-linear/tri-linear Cohesive Zone Modeling." *Composite Structures* 161:505-512.

[22] ASTM E976 - 10. 2010. "Standard guide for determining the reproducibility of acoustic emission sensor response." West Conshohocken, PA: ASTM International.

[23] Wee, K.Y., Koh, S.K. and Lee, J.H. 2002. "Mode I Fracture Resistance Characteristics of Graphite/Epoxy Laminated Composites." *Polymer Composites* 21(2):155-164.

[24] de Moura, M.F.S.F., Campilho, R.D.S.G., Amaro, A.M., and Reis, P.N.B. 2010. "Interlaminar and intralaminar fracture characterization of composites under mode I loading." *Composite Structures* 92:144-149.

[25] Pereira, A.B., and de Morais, A.B. 2004. "Mode I interlaminar fracture of carbon/epoxy multidirectional laminates." *Composites Science and Technology* 64:2261-2270.

[26] Tamuzs, V., Tarasovs S., and Vilks, U. 2001. "Progressive delamination and fiber bridging modeling in double cantilever beam composite specimens." *Composites Science and Technology* 68:513-525.

[27] Fotouhi, M., Heidary, H., Ahmadi, M., and Pashmforoush, F. 2012. "Characterization of composite materials damage under quasi-static three-point bending test using wavelet and fuzzy C-means clustering." *Journal of Composite Materials* 46(15):1795-1808.

[28] Pashmforoush, F., Fotouhi, M., and Ahmadi, M. 2012. "Damage characterization of glass/epoxy composite under three-point bending test using acoustic emission technique." *Journal of Materials Engineering and Performance* 21:1380-1390.

[29] Fotouhi, M., Saeedifar, M., Sadeghi, S., Ahmadi Najafabadi, M., and Minak, G. 2015. "Investigation of the damage mechanisms for mode I delamination growth in foam core sandwich composites using acoustic emission." *Structural Health Monitoring* 14(3):265-280.

[30] Soman, K.P., and Ramachandran, K.I. 2004. "Insight into wavelets from theory to practice." India: PrenticeHall.

[31] Wojtaszczyk, P. 1997. "A mathematical introduction to wavelets." London: Cambridge University Press.

[32] Refahi Oskouei, A., Ahmadi, M., and Hajikhani, M. 2009. "Wavelet-based acoustic emission characterization of damage mechanism in composite materials under mode I delamination at different interfaces." *Express Polymer Letter* 3: 804-813.

[33] Francisco, A.T., and de Carvalho, F. 2007. "Fuzzy C-means clustering methods for symbolic interval data." *Pattern Recognition Letters* 28(4) 423-437.

[34] Sadaaki, M. 2003. "Information clustering based on fuzzy multisets." *Information Processing and Management* 39(2):95-213.

In: Focus on Acoustic Emission Research ISBN: 978-1-53611-848-3
Editors: C. Barile and G. Pappalettera

Chapter 4

ACOUSTIC EMISSION RESEARCH ON PLANT FIBRE COMPOSITES

Carlo Santulli*
Università degli Studi di Camerino,
School of Architecture and Design, Ascoli Piceno, Italy

ABSTRACT

This work concerns the use of acoustic emission to monitor in real time the behaviour of composites reinforced with fibres of vegetable origin. Studies on these materials are gradually growing during last decades, especially in that they may be recommended with respect to fibreglass to achieve improved sustainability and reduced weight. This trend becomes increasingly of interest whenever other aspects linked with environmental protection come into the picture, such as the use of agrowaste and/or the development of biodegradable matrices. However, plant fibre composites still show significant limits, in terms of mechanical performance, durability and geometrical tolerance. This last aspect is particularly critical due to the presence of significant amounts of porosities in their structure, owed primarily to the fact that vegetable fibres are hierarchical structures, in which defects accumulate at different levels, from the microscopic to the macroscopic ones. In this scenario, acoustic emission can play a role, since it allows comparisons between materials in the same testing situation and does not introduce further damage into the structure. On the other side, a review of the studies

* E-mail: carlo.santulli@unicam.it.

performed so far on plant fibre composites using acoustic emission may have some significance, because these involve many different fibres in various loading situations. As a consequence, summarising the main achievements can give some hints on the possibilities for AE in this field in the next future, unveiling the conditions that may present interest for further studies.

INTRODUCTION

An option for improved sustainability of composites is the replacement of fibreglass with plant fibres, with different aims: an initial consideration would concern the reduction of weight, since vegetable fibres, being based on cellulose, do not exceed a 1.5-1.6 specific weight in the worst case, against around 2.5-2.7 for glass fibres. On the other side, beyond the bare consideration of reduced weight, the need of making an appropriate fibre selection is crucial to the success of the operation. In practice, the above mentioned replacement implies passing from one type of fibre produced industrially and therefore with controllable quality, with a large number of possible fibres, extracted from plants by processes that are not normally completely controlled by the end users, especially because they are not intended for application in engineering materials in the first place [1]. In particular, there are factors that have a very large influence on fibre characteristics, but on the other side are preliminary to the process of fibre production, such as e.g., the agronomical conditions of growth of the vegetable species. Some fibres, such as nettle, which might be in general terms available for composite production via the development of textiles adapted for matrix impregnation, present a very large variability in terms of mechanical properties, an issue which is not totally resolved in terms of crop management [2]. One of the consequences of dimensional instability of plant fibres is that it is very difficult to obtain rather constant properties after repeated testing, therefore in prospected service. On the other side, treatment with chemicals, such as (with growing levels of refinement in terms of surface action) mercerization with sodium hydroxide [3], acetilation, silanization [4] or compatibilisation using maleic anhydride, is able to improve their performance [5].

Further reflection would also indicate the possibility to exploit by-products from other manufacturing sectors: these would also offer significant potential in the production of weave structures tailored for use in composites:

however, this possibility has been only quite scantily exploited so far [6]. A good example of this practice would be adopting for fibre production methods employed in sectors, such as the textile one. On the other side, these methods cannot always be considered fully sustainable: for example, they involve significant water consumption for fibre stripping. It needs to be considered in fact that fibres are in most cases arranged as helical coils into vegetable structures [7], while in contrast engineering applications would require long stretches of unidirectional fibres, with the added quality to be able to withstand significant torsion moments, if ultimately textile products, such as plain, twill or satin weave, are to be produced.

Fibre stripping is normally followed once again by extensive treatment, usually with chemicals, such as sodium hydroxide, to obtain fibres of sufficient strength and rigidity to be adapted for use. This has been examined also in terms of life cycle analysis (LCA), to which for example the automotive industry is very interested, before starting with an extensive production of panels or other semi-structural components involving plant fibre composites [8]. This leads to the general necessity to improve the quality of fibre products, at the same time limiting the amount of chemicals introduced, to provide ultimately composites with a sufficient performance for service. In this context, acoustic emission can once again play a role, being particularly sensitive to inelastic deformation, a fundamental aspect to ensure the suitability for use of plant fibre composites, which was investigated for example in the case of flax/unsaturated polyester laminates [9].

There are also other properties that have received limited attention, but are interesting nonetheless, such as the capacity of plant fibre composites to show pronounced vibration damping: this specific characteristic can be further improved by the addition of glycerol and polyglycerol, e.g., to epoxy/flax composites, as suggested by acoustic testing [10]. Studies in this field allowed elucidating the fact that the mode of fracture of plant fibre composites is distinctively different from what is experimented with fibreglass. In particular, it was observed that damage progression involves first kink bands straightening when fibres are first loaded, which causes partial debonding at the fibre-matrix interface. Further cyclic loading normally results in extensive friction between microfibrils within the kink bands during their twisting and untwisting motion [11]. It can be interesting therefore noticing that acoustic emission monitoring is able to discern between these different mechanisms, as it is often the case with other composites, such as fibreglass or carbon fibre composites. Also in the case of plant fibre composites, AE amplitude levels are considered significant to discern between matrix-led damage events, such as matrix cracking and delamination, and fibre-led ones, such as pull-out and fibre breakage, revealed by lower or higher amplitudes, respectively [12]. Another possibility to improve this classification is correlating AE amplitude signals with microscopic observations. On hemp fabric/epoxy composites, this allowed identifying amplitude and tracking, during tensile tests, matrix cracking, interfacial debonding and reinforcement fracture, typically by their different amplitude [13]. The different damage modes were also discerned according to their dissimilar AE frequency distribution, in a work on sisal fibre composites: it was also clarified that an improved tensile and flexural performance of the materials would be obtained by filling with a higher amount of epoxy resin the fibre lumens by controlling in a tailored way injection direction and injection pressure [14]. More recently, the emphasis put on the use of biodegradable polymers, such as for example poly(lactic acid) (PLA) or polycaprolactone (PCL), would offer even more possibilities to the use of plant fibres as their reinforcement, since these are naturally compatible with biopolymers. In particular, acoustic emission proved useful e.g., for the assessment of seawater effect on flax/PLA composites: here, the growing effect of ageing and temperature was evidently perceivable by the gradually rising AE amplitudes (Figure 1) [15].

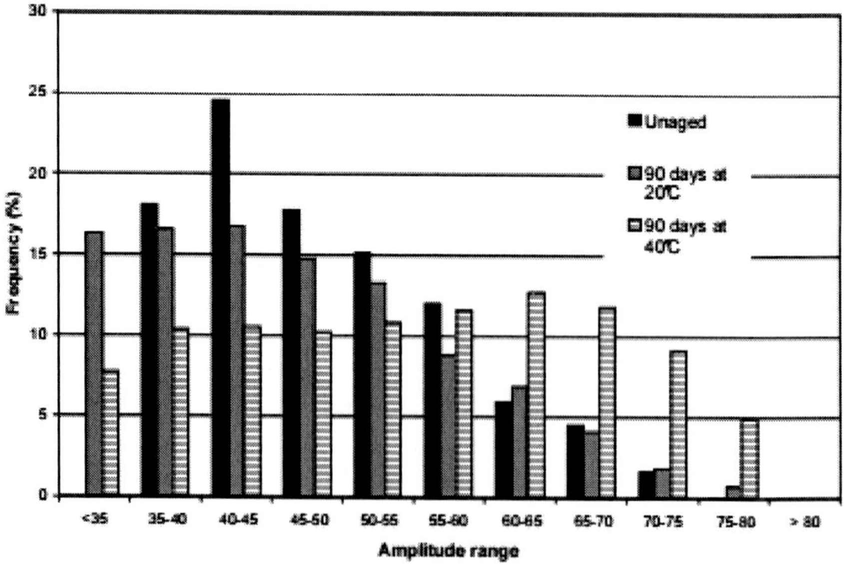

Figure 1. Comparison of distributions of amplitudes of AE signals during tensile tests on samples unaged and after aging at 20 and 40°C for 90 days [15].

In general terms, monitoring the mechanical behaviour of composites, such as fibreglass, using acoustic emission is a method, which has been widely applied in literature: this led in some cases to an appreciable and possibly reliable distinction between the different damage modes [16]. Passing to plant fibre composites, these are perceived in the first instance as a replacement, intended as more environmentally sustainable, for fibreglass and therefore preferably designed in ways as close as possible to this, although geometrical tolerances are obviously much larger, as far as the reinforcement architecture is involved [17]. This description purposely excludes glass/plant fibre hybrids, whose AE signature has nevertheless received some attention, but in which the amount of plant fibres introduced is often limited and furthermore most AE is detected inside the fibreglass layers [18-19]. Dealing with plant fibre composites, a significant problem lies in the fact that the level of acoustic emission detected is much lower, since ligno-cellulosic fibres are naturally more dampened than glass ones. In other words, the interest of acoustic emission monitoring in these materials is significantly linked to the type and characteristics of loading applied to the above. Despite what can appear as a limitation in the above discussion, it is worth noting that some production (or pre-production) quality control was nonetheless performed on plant fibre composites using AE. For example, the quality of flax slivers was inspected using acoustic emission prior to their possible introduction in a composite, which, coupled with further characterization, provided sufficient information to proceed to fibre selection [20]. This study interestingly proposes the investigation of possible damage modes in flax fibre bundles and the specific characteristics of the acoustic emission signals associated to each of these, starting from the consideration that the onset of different damage modes is not easily revealed by the bare observation of mechanical testing diagrams. This has been also demonstrated in the case of flax technical fibres, according to AE amplitude classification reported in Figure 2.

Hypotheses on the micro-failure mode of composites including plant fibres have been also formulated, with the assistance of acoustic emission, in cases where the reinforcement is obtained by using less diffuse fibres, such as *Phormium tenax* (New Zealand flax) [21]. More specifically, it appeared that the scattered distribution of event energies could be attributed to the diversity of types of technical fibres, which go from assemblies of a few thin-walled cells to assemblies of hundreds of thick-walled fibre cells [22]. The sequence of failure events in plant fibre composites, as regards fibres, is also normally characteristic, as it can be observed in Figure 3, referring to flax, which was indicated to be linked to the onset of AE signals of given characteristics [23].

Proposed damage mechanisms from the correlation of the loading curve and the acoustic emission. (N/% def. means newton per percent deformation).

Damage mechanisms	Analysis of the loading curve		Analysis of the acoustic signals			Different families of damage mechanisms
	Reduction of the sliver rigidity (N/% def.)	Elastic energy variation	Hit duration (μs)	Absolute hit energy (10^9 ato J)	Amplitude (dB)	
Friction among ultimate fibres or fibre bundles	Non detectable	Non detectable		Low energy	Low	#1
Rupture of ultimate fibres	Very low	Non detectable		Low	Medium	
Delamination among or within fibre bundles	<0.8	Low	>1500	<5	<80	#2
Partial ruptures of fibre bundles	<0.8	Low/Medium	>1500	>5	>80	
Catastrophic rupture of the fibre bundles	>0.8	High	>1500	>5	80–100	#3

Figure 2. Damage mechanisms correlating loading curve and acoustic emission [20].

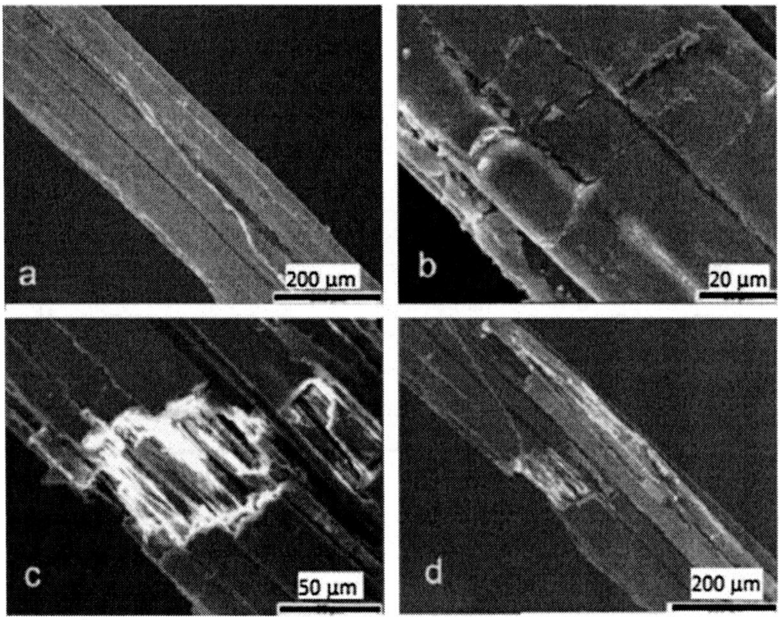

Figure 3. Failure sequence in a technical flax fibre: (a) axial (longitudinal) debonding and fibrillation along the elementary fibres (AE amplitude <35 dB); (b) radial (transverse) cracking in the elementary fibres (amplified effect attributed to stress concentration) (AE amplitude 35–60 dB); (c) "tearing-type" fracture within and through the elementary fibres (AE amplitude 35–60 dB); (d) fracture completed by fracture of the elementary fibres and their constituting micro-fibrils (AE amplitude >60 dB) [23].

AE IN PLANT FIBRE COMPOSITES

General Considerations

A previous review on the application of acoustic emission (AE) to monitoring of plant fibre composites suggested that on these materials AE was a suitable method for the identification of failure modes and other quite confined situations, for example interface studies trying to elucidate adhesion between matrix and fibres [24]. In the few years that have passed from the above review, the situation has slightly evolved, notably on some aspects, such as a larger interest to studies that involve environmental degradation, recognising that this is often a limit to the success of plant fibre composites.

Table 1. Fibres most commonly used in plant fibre composites

Plant	Botanic name	Extracted from ...
Abaca	*Musa textilis*	Leaf
Bamboo	Various species	Stem
Banana	*Musa sapientum*	Leaf
Celery	*Apium graveolens*	Stem
Coir	*Cocos nucifera*	Fruit
Cotton lint	*Gossypium* sp.	Seed hair
Curaua	*Ananas erectifolius*	Leaf
Date palm	*Phoenix dactylifera*	Leaf
Esparto	*Stipa tenacissima/Lygeum spartum*	Stem
Flax	*Linum usitatissimum*	Stem
Hemp	*Cannabis sativa*	Stem
Henequen	*Agave fourcroydes*	Leaf
Indian grass	*Sorghastrum nutans*	Stem
Jute	*Corchorus* sp.	Stem
Kapok	*Ceiba pentandra/acuminata*	Fruit hair
Kenaf	*Hibiscus cannabinus*	Stem
Nettle	*Urtica dioica*	Stem
Oil palm	*Elaeis guineensis*	Fruit
Okra	*Abelmoschus esculentus*	Stem
Palmyra/Borassus	*Borassus flabellifer*	Leaf base
Phormium (New zealand flax)	*Phormium tenax*	Leaf
Piassava	*Attalea funifera*	Leaf/Leaf base
Pineapple	*Ananas comosus*	Leaf
Ramie	*Boehmeria nivea*	Stem
Roselle	*Hibiscus sabdariffa*	Stem
Sansevieria	*Sansevieria cylindrica*	Leaf
Sisal	*Agave sisalana*	Leaf
Spanish broom	*Spartium junceum*	Stem
Sunn hemp	*Crotalaria juncea*	Stem
Switchgrass	*Panicum virgatum*	Stem
Tamarind	*Tamarindus indica*	Fruit
Vetiver	*Vetiveria zizanoides*	Stem

In the following considerations, indications are reported of the cases where AE inspection of plant fibre composites proved successful and worth to be pursued to contribute to the explanation of the material behaviour. The potential of acoustic emission in this kind of inspection can be summarised first of all in terms of its possible contribution to a rational selection of the fibre to be used. Despite extensive research, most works concern just a few well diffused fibres, namely flax, hemp, jute, sisal and some other ones, but the possibilities for selection are very large, at least theoretically, arriving to potentially include many more of them, extracted from bast, leaf or seed from no less than a few dozens of botanical species at least: a tentative list is offered in Table 1. On the other side, it has been suggested that most applications in the design field would involve flax and hemp fibre composites, which are the most developed in terms of production so far (for reasons, such as e.g., availability of preforms) [25].

It is obvious at this point that rational fibre selection would benefit from carrying out comparative studies on plant fibre composites with different fibres, studies to which acoustic emission can give some contribution. For example, a study compares jute, flax and abaca as the reinforcement of polypropylene matrix and the effect of maleic anhydride compatibilisation of the polymer (MAPP) to ensure sufficient interface strength with the fibres. The results show merits and limits of each of the fibres, for example jute fibre composites present higher tensile properties and flexural stiffness, whereas abaca fibre ones are superior in terms of flexural strength: this indicates that an effective selection of the fibre would be strongly dependent from the proposed application [26]. Another study compares interfacial strength of jute and hemp fibres in MAPP compatibilised matrix making use of acoustic emission and comparing its results with interfacial shear strength results, an aspect that will be dealt with, going into some more detail, further down [27].

Another possibility, which is of paramount importance to understand the real potential of plant fibre composites, is the investigation over the maximum amount of fibres that can be introduced in the matrix to ensure effective impregnation. A number of studies exist on this aspect, also in view of the prospected application of biodegradable matrices e.g., oil-based ones [28]. Some of these involve the use of acoustic emission: in [29], birch wood fibres are introduced in different types of polyethylene matrices in amounts from 10 to 40 wt.%, examining the different modes of damage during flexural loading through AE monitoring and proving that the action of maleic anhydride coupling agent reduced the participation of decohesion to laminates failure. In a similar study [30], trend of AE cumulative energy (or counts) curves is compared with that of tensile loading curves: this is a practice quite diffuse in traditional composites, also in the presence of complex reinforcements, such as for example in commingled yarn glass fibre composites [31], and that proved effective also in composites with lignocellulosic reinforcement. In this particular case, it was suggested by AE that exclusive damage modes, such as decohesion and matrix/fibre friction, increased the energy measured during material degradation.

Quite obviously and in general terms, therefore assuming that fibres are as evenly distributed as possible in the material, the higher the amount of fibres, the better will be the performance of the composite. On the other side, since plant fibres, in contrast with glass fibres, have not a circular section and are arranged in a hierarchical structure, therefore being formed by filaments, it is unlikely that it is possible to arrange an equal (not to say a higher) amount of fibres as it would be the case with glass fibres. However, the limit is in the aforementioned effective impregnation of fibres by the matrix and this can be checked using acoustic emission, especially in terms of adoption of the optimal mode for composites production [32].

Effect of Treatment and Environmental Degradation

The effect of treatment to provide an improved interfacial strength of the fibres to the matrix in aggressive conditions, such as boiling water tests, in the case of jute fibre composites, either untreated or subjected to alkali or silane treatment, has also been investigated. This has been considered together with the effect of environmental degradation, because in both cases fibres are modified and so is the adhesive shear strength of their interface with the matrix.

As regards acoustic emission, it was found that AE energy and amplitude data are able to provide some information, especially in terms of comparison with results of interfacial shear strength tests (IFSS). In particular, after boiling water test, a larger number of AE events occurred than before and AE energy became lower, which was attributed to microfailure process of swollen fibrils after boiling water test. AE energy, amplitude and the number of AE events also increased by surface treatment, i.e., alkaline and silane, on jute fibres, as the consequence of improved IFSS and tensile strength of jute fibre composites. The result after boiling water test of silane treated jute fibre shows some difference compared with other two cases, since the number of AE events did not show any substantial increase, whereas AE energy and amplitude were higher after boiling water case, which was ascribed to IFSS increase, as well as to enhanced cohesion between fibril and fibril by silane treatment [33]. This can be observed in Figure 4. To evaluate the difference between acoustic emission in the whole fibre and in fibrils, in the case of ramie and kenaf fibre/epoxy composites, different microfailure modes, axial debonding and fibril fracture, coming from fibre bundles and single fibre composites (SFC), respectively, were observed under tension and compression and evaluated both optically and determined non-destructively by AE and FFT analysis [34]. Also in the case of flax fibre composites, AE indicated that matrix–fibres interface weakening is the main damage mechanism induced by water ageing not differently from what happens for fibreglass. Here, when water ageing leads to severe weakening by wetting of the composites, during tensile loading of flax/epoxy composites AE activities attributed to matrix cracking are reduced, while those linked with debonding basically do not appear before failure [35]. In a similar study on hygrothermal ageing of flax/epoxy composites exposed into an environmental test chamber with a relative humidity (RH) of 90% at different temperatures (20 and 40°C), the acoustic signals recorded during tensile tests were analysed according to amplitude and duration signal and results are exposed in Figure 5 [36]. Focusing on amplitude frequencies, all specimens have a maximum percentage of the number of hits centred around 40–45 dB, usually considered to be characteristic of matrix damage, such as deformation, tearing and cracking [24]. In contrast, AE amplitudes higher than 55 dB are more significant with ageing and the higher temperature of 40°C makes this shift towards higher amplitudes even more important. Focusing on duration frequencies, the signal duration increases with ageing due to relative humidity only, while the temperature does not appear to have effect on it. AE multivariate analysis on hemp/polypropylene composites with different amounts of fibres (10 to 40 wt. %) and subjected to hygrothermal ageing by a relative humidity of 80% at different temperatures (25 and 50°C) revealed a low number of AE signals induced by matrix-fibre interface failure, for both aged and unaged composites, which was attributed to the ineffective hemp fibres treatment [37]. However, the result from the classification can be used as an indicator to evaluate the quality of fibre-matrix interface, by focusing on the number of AE signals induced by the interface failure and on their amplitude ranges.

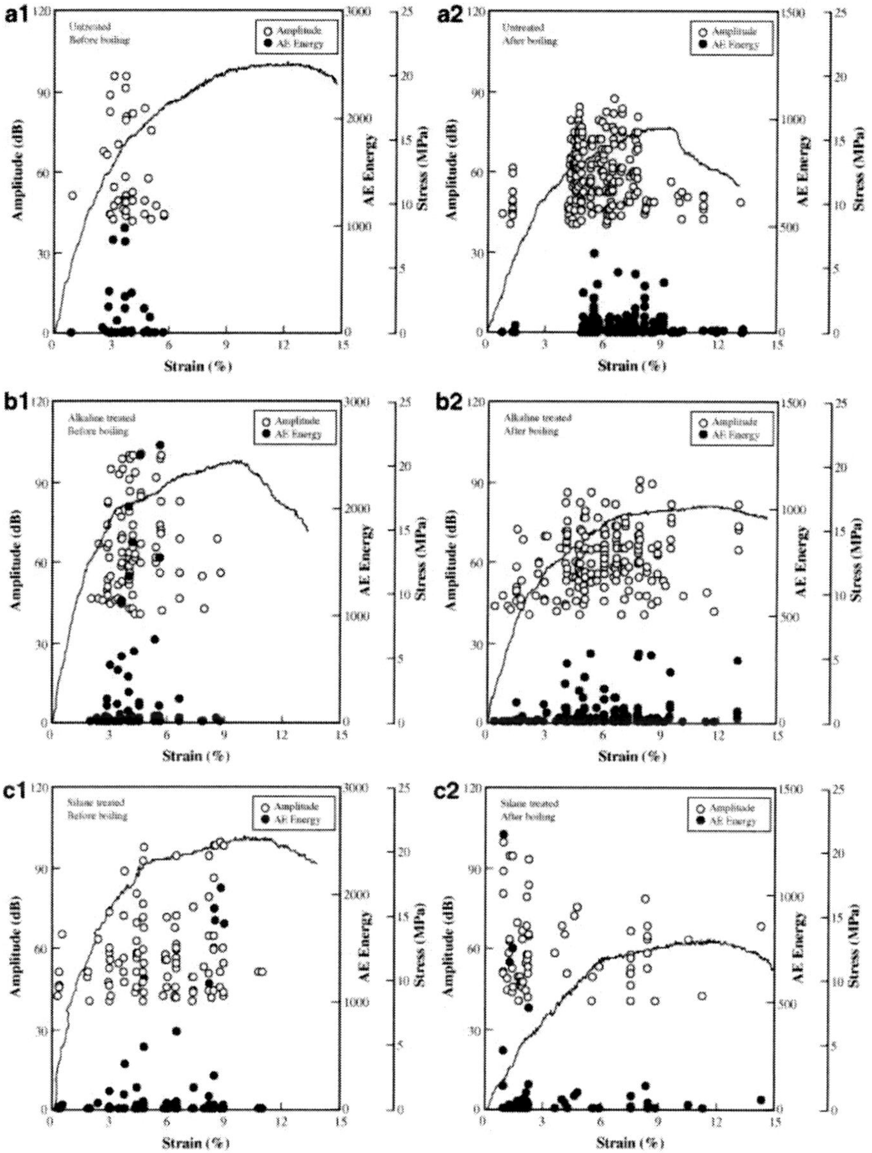

Figure 4. AE amplitude and AE energy of jute fibres/PP composites for: (a) the untreated; (b) alkaline treated; and (c) silane treated [33].

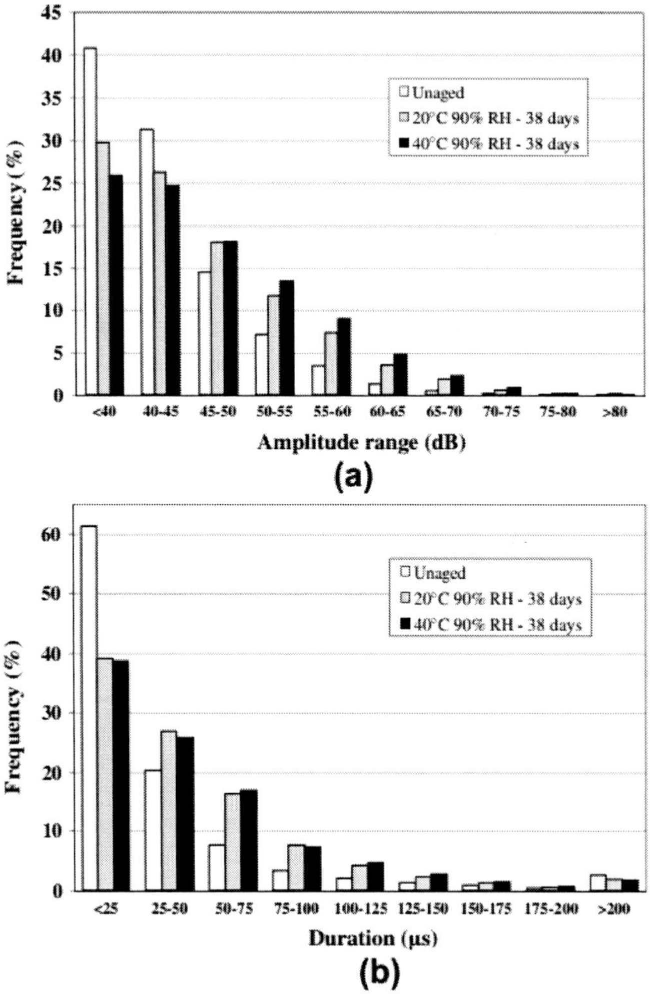

Figure 5. Comparison of distribution of AE signals during tensile tests on unidirectional flax/epoxy composites (unaged and after 38 days of ageing) according to: amplitudes (a) and duration signal (b). Amplitude (a) and duration distribution [36].

Performance Degradation as the Consequence of Mechanical Damage

Falling weight impact (IFW) can be intended as a quite reliable method to simulate accidental damage occurring during service and is therefore widely applied for materials that are supposed to be used in fields, such as automotive, where the rate of occurrence of this kind of issues is supposed not to be negligible. It is no surprise thus that also plant fibre composites were subjected to IFW tests and, following this, to post-impact performance evaluation, in some cases monitored by acoustic emission.

In particular, in post-impact cyclic flexural tests on hemp fibre/unsaturated polyester composites it was noticed that AE events' distribution gradually shifts towards higher amplitudes and lower durations with growing impact energy (12-16-20 J). This was explained by substantial damage already taking place during impact, as reflected by high amplitudes, while more serious cracks are opening in the material as the effect of cyclical loading [38]. In the case of jute/unsaturated polyester subjected to post-impact three-point flexural testing, AE revealed instead delamination as the main fracture mode [39].

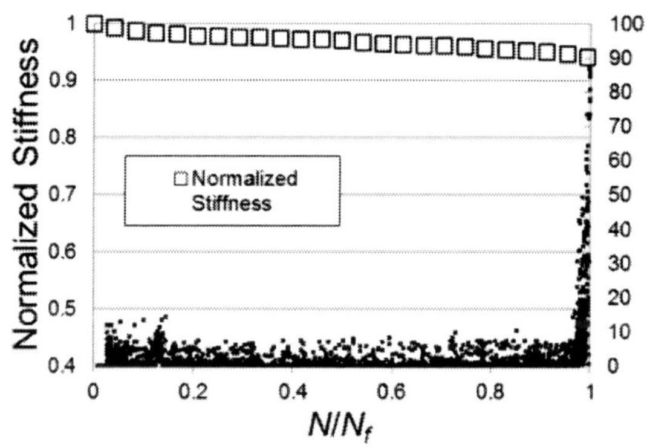

(a) UD Composites

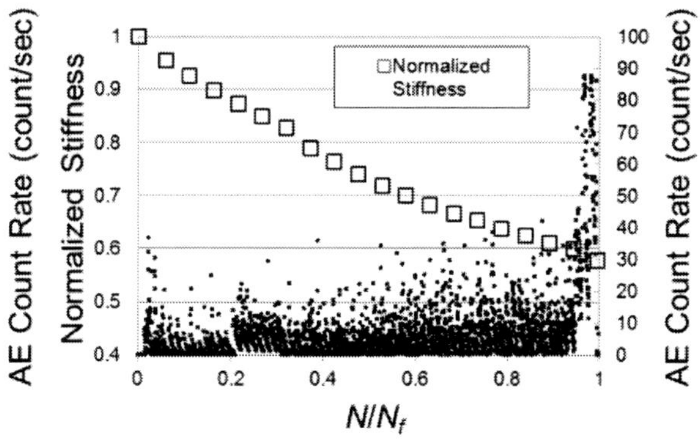

(b) Textile Composites

Figure 6. Normalized stiffness and AE count rate as a function of normalized number of cycles [43].

Studying tensile-tensile fatigue of woven hemp fibre epoxy composites, AE results were compared with the evidences of damage from X-ray computed micro-tomography, indicating a particular similarity [40]. More specifically, the tensile behaviour involved three phases: a significant initial slope at the beginning of the curves, related to transverse (through-thickness) matrix cracks and initial fibre breakages; an almost constant slope, related to the period during which the matrix cracks may couple together and grow, especially along interfaces; the slope gradually increases again just before failure, where from initial fibre fractures an evolution towards strand failures is shown [41]. This work highlights that damage localisation, hence measurement of its extent and position in the material, is possible through acoustic emission: as a matter of fact, position-resolved AE enabled the detection of crack propagation in single edge notched tensile (SEN-T) loaded specimens, while also characterising the failure [42]. Moreover, a clear correlation was also revealed between normalised stiffness of poly(lactic acid)/hemp fibre composites and AE count rate (count/second) during cyclic tensile loading, allowing also distinguishing between damage obtained in composites with unidirectional hemp and in plain woven textiles [43] (Figure 6). In another work on thermoplastic starch matrix/flax fibre composites, attempting to measure the J-R curve, it was also found that, by weighting and smoothing the located cumulative AE amplitudes, the crack path, which was not clearly observed by visual methods, can be correctly reconstructed [44]. A similar dependence of the work of fracture along the crack, hence with damage modes, such as matrix deformation, fibre pull-out and fibre breakage, with amplitude distribution, was also observed in kenaf fibre/thermoplastic natural rubber/polypropylene composites: in this case, probably due to the matrix employed, the presence of a larger amount of kenaf fibres led to stronger AE signals [45].

CONCLUSION

The application of acoustic emission to the monitoring of behaviour of plant fibre composites, hence reinforced with lignocellulosic fibres of vegetable origin, proved successful in a number of cases, despite the complexity of these materials, due to a number of aspects, which will be briefly exposed below. First of all, introducing plant fibres means that the material is more dampened than traditional composites, such as carbon fibre composites or fibreglass, which in principle should affect AE signal levels and therefore possibly reduce the sensitivity of this technique. This trend has been noticed, but appears strongly matrix-dependent and therefore not necessarily encountered in practice. Further considerations would concern the fact that in plant fibre composites fibre failure has different characteristics than in traditional composites, especially due to the fact that vegetable fibres are formed from spiral coiled structures and therefore hierarchized down to the microscopic level. In other words, this means that, in addition to fibre breakage, the occurrence of fibrillation, hence separation of the vegetable fibres into the constituting filaments needs also to be taken into account.

Despite the above limitations, the distinction between the different damage modes, e.g., matrix cracking, delamination/debonding and fibre fracture, during tensile, flexural and fatigue tests, using acoustic emission appears very promising. Studies did allow effectively comparing the results with those obtained from other techniques, such as X-ray microtomography and scanning electron microscopy. This led even to the possibility of ensuring that crack propagation, as it appears from acoustic emission, bears a close similarity with the measurements obtained by techniques such as interfacial shear strength (IFSS). Another aspect that was deeply investigated is the assessment of environmental degradation due for example to the effect of boiling water or of seawater, coupled with that of temperature: here too, acoustic emission led to promising results, not only confined to a go/no go perspective, but even allowing a sounder characterisation of environmentally induced damage.

Parameters that proved more suitable for inspection are in particular AE amplitudes distribution, but also counts rate and energy, especially when compared with loading curves, while the significance of AE events duration with interfacial damage phenomena is confirmed also on these materials. It may be suggested that future studies would increasingly concentrate on localising damage through the time difference of signal arrival to the different sensors, as this is a growing trend also on traditional composites.

REFERENCES

[1] Wambua, P., Ivens, I., Verpoest, I., Natural fibres: can they replace glass in fibre reinforced plastics?, *Composites Science and Technology* 63 (9), 2003, 1259–1264.

[2] Kumar, N., Das, D., Alkali treatment on nettle fibers. Part I: investigation of chemical, structural, physical, and mechanical

characteristics of alkali-treated nettle fibers, *The Journal of the Textile Institute*, Published online: 14 Nov 2016.

[3] Qin, C., Soykeabkaew, N., Xiuyuan, N., Peijs. T., The effect of fibre volume fraction and mercerization on the properties of all-cellulose composites, *Carbohydrate Polymers* 71 (3), 2008, 458–467.

[4] Bledzki, A. K, Gassan, J., Composites reinforced with cellulose based fibres, *Progress in Polymer Science* 24 (2), 1999, 221-274.

[5] Joshi, S. V., Drzal, L., Mohanty, A. K., Arora, S., Are natural fiber composites environmentally superior to glass fiber reinforced composites?, *Composites Part A* 35 (3), 2004, 371-376.

[6] Goutianos, S., Peijs, T., Nystrom, B., Skrifvars, M., Development of flax fibre based textile reinforcements for composite applications, *Applied Composite Materials* 13 (4), 2006, 199–215.

[7] Bakri, B., Eichhorn, S. J., Elastic coils: deformation micromechanics of coir and celery fibres, *Cellulose* 17 (1), 2010, 1-11.

[8] Alves, C., Ferrão, P. M. C., Silva, A. J., Reis, L. G., Freitas, M., Rodrigues, L. B., Alves, D. E., Ecodesign of automotive components making use of natural jute fiber composites, *Journal of Cleaner Production* 18 (4), 2010, 313–327.

[9] Hughes, M., Carpenter, J., Hill, C., Deformation and fracture behaviour of flax fibre reinforced thermosetting polymer matrix composites, *Journal of Materials Science* 42 (7), 2007, 2499-2511.

[10] Le Guen, M.-J., Newman, R. H., Fernyhough, A., Staiger, M. P., Tailoring the vibration damping behaviour of flax fibre-reinforced epoxy composite laminates via polyol additions, *Composites Part A* 67, 2014, 37–43.

[11] Newman, R. H., Battley, M. A., Carpenter, J. E. P., Le Guen, M. J., Energy loss in a unidirectional flax-polyester composite subjected to multiple tensile load–unload cycles, *Journal of Materials Science* 47 (3), 2012, 1164-1170.

[12] Romhány, G., Karger-Kocsis, J., Czigány, T., Tensile fracture and failure behavior of thermoplastic starch with unidirectional and cross-ply flax fiber reinforcements, *Macromolecular Materials and Engineering* 288 (9), 2003, 699-707.

[13] Bonnafous, C., Touchard, F., Chocinski-Arnault, L., Damage mechanisms in hemp-fibre woven fabric composite, and comparison with glass-fibre composite, *Polymers & Polymer Composites* 19 (7), 2011, 543-552.

[14] Li, Y., Ma, H., Shen, Y., Li, Q., Zheng, Z., Effects of resin inside fiber lumen on the mechanical properties of sisal fiber reinforced composites, *Composites Science and Technology* 108, 2015, 32-40.

[15] Le Duigou, A., Davies, P., Baley, C., Seawater ageing of flax/poly(lactic acid) biocomposites, *Polymer Degradation and Stability* 94 (7), 2009, 1151-1162.

[16] Huguet, S., Godin, N., Gaertner, R., Salmon, L., Villard, D., Use of acoustic emission to identify damage modes in glass fibre reinforced polyester, *Composites Science and Technology* 62 (10-11), 2002, 1433-1444.

[17] Koronis, G., Silva, A., Fontul, M., Green composites: A review of adequate materials for automotive applications, *Composites Part B* 44 (1), 2013, 120–127.

[18] Md Akil, H., Santulli, C., Sarasini, F., Tirillò, J., Valente, T., Environmental effects on the mechanical behaviour of pultruded jute/glass fibre-reinforced polyester hybrid composites, *Composites Science and Technology* 94, 2014, 62-70.

[19] Jawaid, M., Abdul Khalil, H. P. S., Cellulosic/synthetic fibre reinforced polymer hybrid composites: a review, *Carbohydrate Polymers* 86, 2011, 1-18.

[20] Barbulée, A., Jernot, J.-P., Bréard, J., Gomina, M., Damage to flax fibre slivers under monotonic uniaxial tensile loading, *Composites Part A* 64, 2014, 107-114.

[21] De Rosa, I. M., Santulli, C., Sarasini, F., Mechanical and thermal characterization of epoxy composites reinforced with random and quasi-unidirectional untreated Phormium tenax leaf fibers, *Materials & Design* 31 (5), 2010, 2397–2405.

[22] Newman, R. H., Le Guen, M. J., Battley, M. A., Carpenter, J. E. P., Failure mechanisms in composites reinforced with unidirectional Phormium leaf fibre, *Composites Part A* 41 (3), 2010, 353-359.

[23] Romhany, G., Karger-Kocsis, J., Czigany, T., Tensile fracture and failure behaviour of technical flax fibres, *Journal of Applied Polymer Science* 90 (13), 2003, 3638–3645.

[24] De Rosa, I. M., Santulli, C., Sarasini, F., Acoustic emission for monitoring the mechanical behaviour of natural fibre composites: A literature review, *Composites Part A* 40 (9), 2009, 1456-1469.

[25] Pil, L., Bensadoun, F., Pariset, J., Verpoest, I., Why are designers fascinated by flax and hemp fibre composites?, *Composites Part A* 83, 2016, 193–205.

[26] Bledzki, A. K., Mamun, A. A., Faruk, O., Abaca fibre reinforced PP composites and comparison with jute and flax fibre PP composites, *Express Polymer Letters* 1 (11), 2007, 755–762.
[27] Park, J.-M., Quang, S. T., Hwang, B. S., DeVries, K. L., Interfacial evaluation of modified jute and hemp fibers/polypropylene (PP)-maleic anhydride polypropylene copolymers (PP-MAPP) composites using micromechanical technique and nondestructive acoustic emission, *Composites Science and Technology* 66 (15), 2006, 2686–2699.
[28] O'Donnell, A., Dweib, M. A., Wool, R. P., Natural fiber composites with plant oil-based resin, *Composites Science and Technology* 64 (9), 2004, 1135-1145.
[29] Bravo, A., Toubal, L., Koffi, D., Erchiqui, F., Development of novel green and biocomposite materials: Tensile and flexural properties and damage analysis using acoustic emission, *Materials & Design* 66A (5), 2015, 16-28.
[30] Bravo, A., Toubal, L., Koffi, D., Erchiqui, F., Characterization of tensile damage for a short birch fiber-reinforced polyethylene composite with acoustic emission, *International Journal of Materials Science* 3 (3), 2013, 79-89.
[31] Lariviere, D., Krawczak, P., Tiberi, C., Lucas, P., Interfacial properties in commingled yarn thermoplastic composites. Part I: Characterization of the fiber/matrix adhesion, *Polymer Composites* 25 (6), 2004, 577-588.
[32] Kobayashi, S., Takada, K., Nakamura, R., Processing and characterization of hemp fiber textile composites with micro-braiding technique, *Composites Part A* 59, 2014, 1-8.
[33] Park, J.-M., Kim, P.-G., Jang, J.-H., Wang, Z., Hwang, B.-S., DeVries, K. L., Interfacial evaluation and durability of modified jute fibers/ polypropylene (PP) composites using micromechanical test and acoustic emission, *Composites Part B* 39 (6), 2008, 1042-1061.
[34] Park, J. M., Son, T. Q., Jung, J. G., Hwang, B. S., Interfacial evaluation of single ramie and kenaf fiber/epoxy resin composites using micromechanical test and nondestructive acoustic emission, *Composite Interfaces* 13 (2-3), 2006, 105-129.
[35] Assarar, M., Scida, D., El Mahi, A., Poilâne, C., Ayad, R., Influence of water ageing on mechanical properties and damage events of two reinforced composite materials: flax–fibres and glass–fibres, *Materials & Design* 32 (2), 2011, 788-795.

[36] Scida, D., Assarar, M., Poilâne, C., Ayad, R., Influence of hygrothermal ageing on the damage mechanisms of flax-fibre reinforced epoxy composite, *Composites: Part B* 48, 2013, 51–58.

[37] Assarar, M., Scida, D., Zouari, W., Saidane, E. H., Ayad, R., Acoustic emission characterization of damage in short hemp-fiber-reinforced polypropylene composites, *Polymer Composites* 37 (4), 2016, 1101-1112.

[38] De Rosa, I. M., Dhakal, H. N., Santulli, C., Sarasini, F., Zhang, Z. Y., Post-impact static and cyclic flexural characterisation of hemp fibre reinforced laminates, *Composites Part B* 43 (3), 2012, 1382-1396.

[39] Dhakal, H. N., Arumugam, V., Aswinraj, A., Santulli, C., Zhang, Z. Y., Lopez-Arraiza, A., Influence of temperature and impact velocity on the impact response of jute/UP composites, *Polymer Testing* 35, 2014, 10–19.

[40] De Vasconcellos, D. S., Sarasini, F., Touchard, F., Chocinski-Arnault, L., Pucci, M., Santulli, C., Tirillò, J., Iannace, S., Sorrentino, L., Influence of low velocity impact on fatigue behaviour of woven hemp fibre reinforced epoxy composites, *Composites Part B* 66, 2014, 46-57.

[41] De Vasconcellos, D. S., Touchard, F., Chocinski-Arnault, L., Tension–tension fatigue behaviour of woven hemp fibre reinforced epoxy composite: A multi-instrumented damage analysis, *International Journal of Fatigue* 59, 2014, 159-169.

[42] Acha, B. A., Marcovich, N. E., Karger-Kocsis, J., Biodegradable jute cloth reinforced thermoplastic copolyester composites: fracture and failure behaviour, *Plastics, Rubber and Composites* 35 (2), 2006, 73-82.

[43] Kobayashi, S., Damage behavior of hemp fiber reinforced poly(l-lactic acid) composites under fatigue loading, *Journal of Solid Mechanics and Materials Engineering* 7 (2), 2013, 317-323.

[44] Romhány, G., Czigány, T., Karger-Kocsis, J., Determination of J–R curves of thermoplastic starch composites containing crossed quasi-unidirectional flax fiber reinforcement, *Composites Science and Technology* 66 (16), 2006, 3179-3187.

[45] Anuar, H., Ahmad, S. H., Rasid, R., Surip, S. N., Czigany, T., Romhany, G., Essential work of fracture and acoustic emission study on TPNR composites reinforced by kenaf fiber, *Journal of Composite Materials 41,* 2007, 3035-3049.

In: Focus on Acoustic Emission Research ISBN: 978-1-53611-848-3
Editors: C. Barile and G. Pappalettera

Chapter 5

ACOUSTIC EMISSION MONITORING OF IN-SERVICE LARGE-SIZED STRUCTURES

Gianni Niccolini, Giuseppe Lacidogna[*] and Alberto Carpinteri
Department of Structural, Geotechnical and Building Engineering,
Politecnico di Torino, Torino, Italy

ABSTRACT

The acoustic emission (AE) monitoring technique has the potential for performing an effective monitoring of the integrity of large-sized structures, like those in Civil and Mechanical Engineering, by means of a small number of sensors. Based on this method, an on-site application of the AE technique was conducted in order to locate the damaged zones in a carbon fiber reinforced concrete beam tested in bending. We used the AE technique also to assess and evaluate the current condition and operational capability of an in-service double girder bridge crane. The AE sensors were applied on one of the two beams of the bridge crane to diagnose and solve possible issues related to micro structural degradation phenomena. In both cases, localization and evolution of possible damage process were characterized from the received AE signals exploiting different inversion methods based on a seismological approach. The final goal will be to perform machine prognosis and calculate life expectancy of these bearing structures.

[*] Corresponding Author: giuseppe.lacidogna@polito.it.

Keywords: acoustic emission, damage assessment, crack location, large-sized structures, real-time monitoring, fracture precursors

1. INTRODUCTION

Fracture in heterogeneous materials is a complex phenomenon which involves a wide range of scales, from microcracking to earthquake ruptures, including structural failures [1 - 4]. In this framework, acoustic emission (AE) monitoring during loading experiments provides a tool for understanding the occurrence of fractures at larger scales [5]. Indeed, this approach has recently provided the opportunity to develop universal scaling laws which reflect the scale-invariance and the self-similarity of fracture processes in time, space and magnitude domains [6 - 12]. These studies should hopefully contribute to solving the main problem of remaining life assessment of structural elements, such as machinery and bearing structures.

The AE monitoring technique is unique among non-destructive evaluation methods as it is a passive technique, unlike ultrasonic testing or radiography where energy is injected into the monitored object for measurements. Indeed, the AE phenomenon is the spontaneous release of stress waves (typically in the form of ultrasound) by growing defects in materials subjected to loading or deformation (Figure 1 top). Therefore, acoustic emissions are analyzed to gain insight into micro-structural damage phenomena such as micro-cracking and anomalous audible noises in structures under service loads (Figure 1 bottom). The AE monitoring offers the possibility of covering large structural elements in one examination with a small number of sensors to locate damage without the need of a priori knowledge of the defective area's position. With this testing method, defects are listened at the moment of their occurrence, allowing remote and real-time monitoring as a warning system to avoid catastrophic structural failures and to produce cost and time savings (Figure 2). Here, two seismological inversion methods to detect and locate evolving damage processes from received AE data are described in the following section and then, they are applied to two case studies of in-service large-sized structures subjected to AE monitoring.

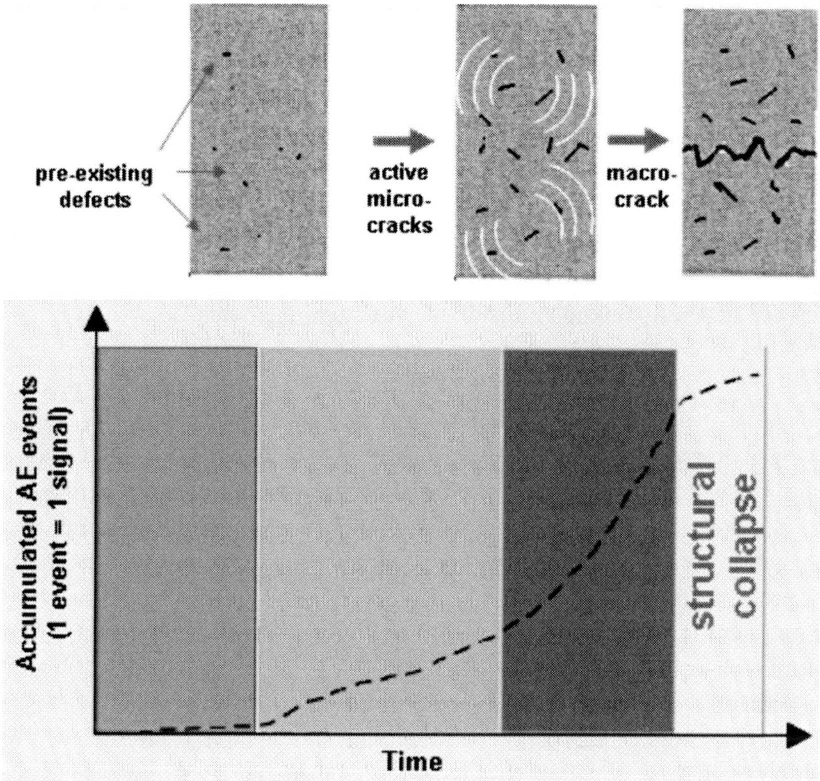

Figure 1. Different stages of an evolving damage process (top), and typical diagram of the corresponding AE activity (bottom).

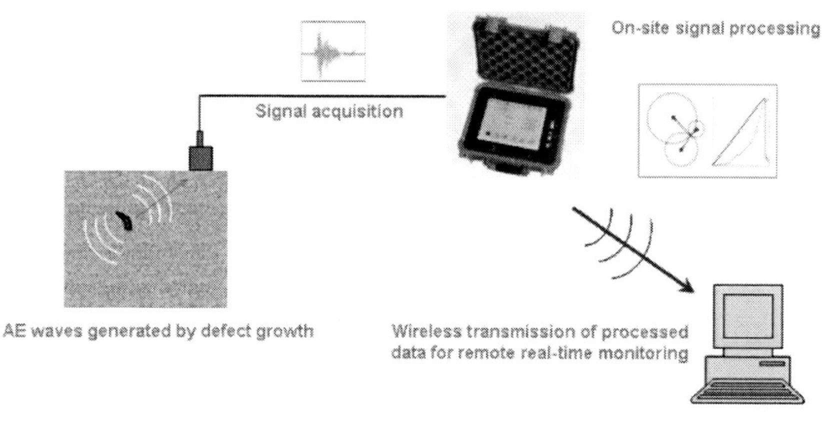

Figure 2. AEmission ® measuring system for microcrack detection.

2. Inversion Methods

2.1. Acoustic Emission Source Location by LSM

Source localization concepts originated in seismology where the aim was to localize the hypocenter or epicenter of an earthquake from seismograms obtained at stations distributed over the Earth's surface [13].

Similar problems are posed when attempting to locate sources of acoustic emission, generally unknown inside the structural elements. The knowledge of the wave propagation characteristics between source and receiver is necessary. For the most general case, these characteristics depend on the mode of propagation, the elastic modulus, the existence of attenuation due to heterogeneities and the anisotropy of material. The equations used for source location in concrete and metals are based on the assumptions of homogeneous and isotropic medium and point-like sources, implying spherical wave propagation. We talk about 'point-like sources' as microcracking is a step-by-step process involving the creation of very small rupture surfaces at a time, each of them is an AE source; due to their limited extent, these source rupture surfaces can be regarded as point sources.

Generally speaking, for single point-like sources in geometries having continuous straight line paths between the source and each receiver, the location technique is called 'triangulation', where the differences in arrival times of AE waves at various sensors are used for source location.

Referring to Figure 3, suppose the AE source is a micro-fracture occurring at (x, y, t) while 3 sensors are applied at positions (x_1, y_1), (x_2, y_2) and (x_3, y_3). The i-th sensor S_i located at (x_i, y_i) will experience the first wave motion when:

$$\sqrt{(x-x_i)^2+(y-y_i)^2} \equiv d_i = v(t_i - t) \qquad (31)$$

being t_i the arrival time of the first wave motion at S_i sensor, and v the wave velocity. If t_i is the arrival time at another sensor S_j, $\Delta t_{ij} \equiv t_i - t_j$ is the measured arrival time difference for the first wave motion recorded at S_i and S_j.

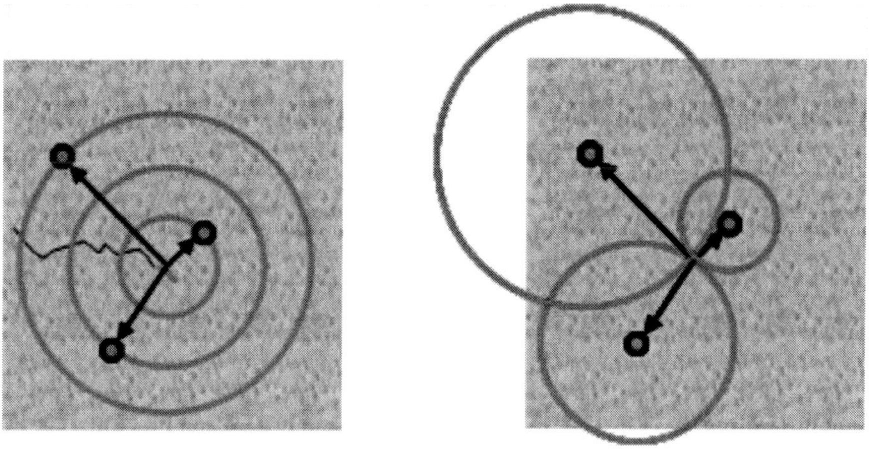

Figure 3. First wave motion generated at time t by an active crack reaching three sensors at different times (left); geometrical meaning of equation.31 (right).

The triangulation process requires the solution of a system of simultaneous equations such as:

$$d_i - d_j = v\,\Delta t_{ij} \tag{32}$$

For the array of n sensors, up to $n-1$ non-linear equations can be formed in this manner. In case of a 2D location problem with three unknowns x, y and v, three measured Δt_{ij} differences, i.e., $n = 4$ detecting sensors, are needed. Note that in the 2D problem of Figure 3, only three detecting sensors are necessary if the wave velocity v is known and the solution is given by linear methods. Actually, equations 32 can be treated with a minimum error approach when v is unknown, as explained below.

The observed arrival time t_i of the first wave motion at the i-th sensor can be considered as being given by an integral along the actual ray path Γ_i (Figure 72 left):

$$t_i = t + \int_{\Gamma_i} \left(ds/v(\boldsymbol{r}) \right) + m_i \tag{33}$$

where m_i is the error affecting the measure t_i, v (**r**) is the wave velocity field in the specimen and t is the origin time of the source crack. Detailed knowledge of the wave propagation characteristics (i.e., Γ_i and v (**r**)) between source and receiver is not possible and then the exact source position will remain unknown and one can only speculate as to the accuracy of the solution given by the triangulation (Figure 4 left).

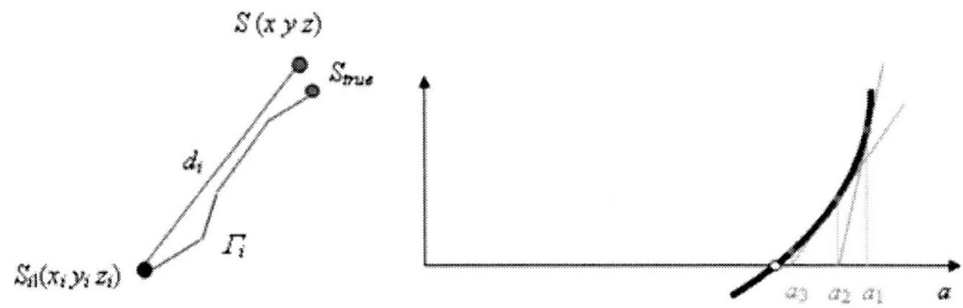

Figure 4. Differences between physical and theoretical (straight) ray paths and between actual and estimate source position (left); geometrical representation of two iterations of Newton's method in a 1-D problem (right).

If t_i is the i-th datum, the theoretical value, T_i given by the triangulation is:

$$T_i = t + d_i / v \tag{34}$$

Since wave propagation along straight ray paths is nothing but an approximation, it will generally result that $t_i \neq T_i$ and so their differences: $\Delta t_{ij} \neq \Delta T_{ij}$. Inserting the explicit expression for ΔT_{ij} into this inequality provides the error or time-residual associated with Eq. (2):

$$e_{ij} = \Delta t_{ij} - \left(d_i - d_j\right) / v \neq 0 \tag{35}$$

The classic approach to AE source location is minimizing a sum of squared residuals and so it is called least squares method (LSM). The idea of applying LSM to source location problems apparently originated with the seismologist L. Geiger [13] to estimate earthquake hypocentral coordinates.

If n sensors have experienced an AE wave, the residuals can be defined according to Eq (5):

$$e_{ij} = \Delta t_{i1} - (d_i - d_1)/v \, , \qquad (36)$$

where the subscript 1 is related to an arbitrarily chosen reference sensor, e.g., the closest one to the source, and $i = 1,\ldots, n$.

The total error for all $n - 1$ equations is simply the sum of all squared time-residuals:

$$X^2 = \sum_{i=2}^{n} e_i^2 \qquad (37)$$

LSM seeks to determine the source location-estimate, which minimizes χ^2. The minimum conditions:

$$\frac{\partial \chi^2}{\partial a_k} = 0 \qquad \text{with } k = 1,\ldots,4 \qquad (38)$$

are solved for $\mathbf{a} \equiv (x, y, z, v)$ in the source location 3D problem. Measured inputs are then the sensor coordinates $(x_1, y_1, z_1), (x_2, y_2, z_2), \ldots, (x_n, y_n, z_n)$ and the arrival time differences Δt_{ij}, while knowledge of v is not required since it can be included among the unknowns.

Nonlinear systems such as that of equation 38 are usually solved by iterative techniques, such as Newton's method.

In Newton's method one starts with a trial guess $\mathbf{a}^{(1)}$ of the solution, then replaces the function $\partial\chi^2/\partial a_k$ by its tangent in $\mathbf{a}^{(1)}$ and computes the zero of this tangent solving a linear system:

$$\frac{\partial \chi^2}{\partial a_k}\left(\mathbf{a}^{(1)}\right) + \left(a_m - a_m^{(1)}\right)\frac{\partial}{\partial a_m}\frac{\partial}{\partial a_k}\chi^2\left(\mathbf{a}^{(1)}\right) = 0 \qquad (39)$$

In equation 39, and elsewhere, when an index variable appears twice in a single term it implies that we are summing over all of its possible values.

This zero of the tangent $\mathbf{a} = \mathbf{a}^{(2)}$ will typically be a better approximation to the function's zero, and the method can be iterated (Figure 72 right). The procedure usually converges to the solution after a few iterations. At the i-th iteration the coefficient matrix of the system is the Hessian matrix:

$$H_{mk}^{(i)} \equiv \frac{\partial^2 \chi^2\left(\mathbf{a}^{(i)}\right)}{\partial a_m \partial a_k} = 2\frac{\partial e_j}{\partial a_m}\left(\mathbf{a}^{(i)}\right)\frac{\partial e_j}{\partial a_k}\left(\mathbf{a}^{(i)}\right) + 2e_j\left(\mathbf{a}^{(i)}\right)\frac{\partial^2 e_j}{\partial a_m \partial a_k}\left(\mathbf{a}^{(i)}\right) \tag{40}$$

Unfortunately, Newton's method needs the computation of second derivatives, which may be inaccurate causing convergence problems. An algorithm that avoids these ill-conditioning problems reduces computational cost and tends to work well when residuals are small and not highly nonlinear is the Gauss-Newton method, usually adopted in Seismology [14, 15].

The Gauss-Newton method makes an approximation to $H_{mk}^{(i)}$ by neglecting the final term in equation 40 containing the second derivatives of residuals to give:

$$H_{mk}^{(i)} \cong 2\frac{\partial e_j}{\partial a_m}\left(\mathbf{a}^{(i)}\right)\frac{\partial e_j}{\partial a_k}\left(\mathbf{a}^{(i)}\right) \tag{41}$$

Therefore, the i-th iteration of the Gauss-Newton method is solving:

$$[\mathbf{G}^{(i)}][\mathbf{G}^{(i)}]^T \boldsymbol{\delta}^{(i)} = -[\mathbf{G}^{(i)}]^T \mathbf{e}^{(i)} \tag{42}$$

where:

$$\left[\mathbf{G}^{(i)}\right] \equiv \begin{bmatrix} \frac{\partial e_2}{\partial a_1}\left(\mathbf{a}^{(i)}\right) & \dots & \frac{\partial e_n}{\partial a_1}\left(\mathbf{a}^{(i)}\right) \\ \dots & & \dots \\ \frac{\partial e_2}{\partial a_4}\left(\mathbf{a}^{(i)}\right) & \dots & \frac{\partial e_n}{\partial a_4}\left(\mathbf{a}^{(i)}\right) \end{bmatrix} \tag{43}$$

is the Jacobian matrix, the columns of which are the gradient of the components of the vector:

$$\mathbf{e}^{(i)} \equiv \begin{bmatrix} e_2\left(a^{(i)}\right) \\ \dots \\ e_n\left(a^{(i)}\right) \end{bmatrix} \tag{44}$$

while $\boldsymbol{\delta}^{(i)} = \mathbf{a}^{(i+1)} - \mathbf{a}^{(i)}$ is the model correction at the i-th iteration.

A further numerical difficulty can arise in the solution of equation 42, since $[\mathbf{G}][\mathbf{G}]^T$ can be a nearly singular matrix, and its computed inverse may be inaccurate still causing convergence problems. Lee and Lahr [16] associated the near singularity with model parameters that are poorly constrained by the data. These problems can be overcome by using techniques based on the Singular Value Decomposition [15] to compute the pseudo-inverse instead of the exact inverse of $[\mathbf{G}][\mathbf{G}]^T$.

2.2. The *b*-value Analysis

In seismology, earthquakes of larger magnitude occur less frequently than ones of smaller magnitude. That has been quantified by a phenomenological law, proposed by Gutenberg and Richter (GR law [1]):

$$\mathrm{Log}_{10} N\left[> M\right] = a - bM \tag{45}$$

Where $N(>M)$ is the cumulative number of earthquakes with magnitude greater than M in specified area and time span, b and a are positive constants varying from region to region. The constant b, often termed as b-value, is the slope of the log-linear frequency-magnitude diagram.

Also the scaling of the amplitude distribution in AE signals are successfully described by the GR law, where $M = \mathrm{Log}\, A_{\mathrm{max}}$ is the magnitude of an AE event with peak amplitude A_{max}.

Damage assessment by AE signals emerging from growing microcracks has been carried out exploiting different statistical methods, such as the analysis of the synthetic parameters β_{t} [17] and b-value. In particular, the b-value decreases systematically with the damage evolution as it presents the relative number of small-magnitude to large-magnitude events [18 - 22]. Indeed, micro-cracking generates mainly a large number of low-magnitude acoustic emissions, while macro-cracking generates also high-amplitude acoustic emissions. Therefore, micro-cracking leads to high b-values (ranging from 1.5 to 2.5), while macro-cracking leads to comparatively lower b-values (tending to 1 or less as impending failure approaches). The previous discussion can be summarized as follows (Figure 5):

$b > 1.7$: microcracks are dominant;
$1.2 < b < 1.7$: macrocracks start to appear;
$1 < b < 1.2$: macrocracks are opening up to the failure.

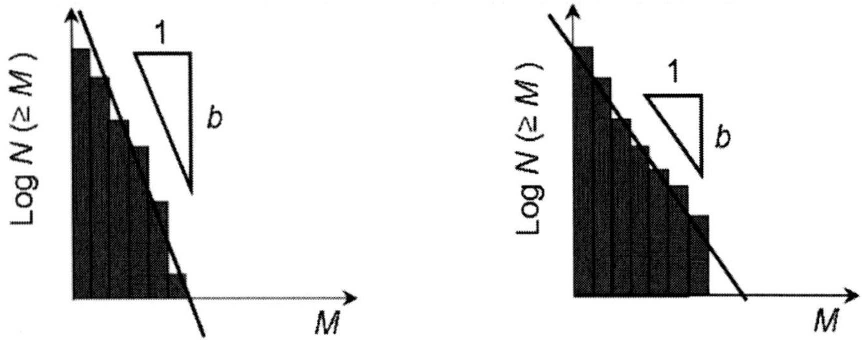

Figure 5. Evolution of GR b-value from microcracking (left) to macrocracking (right).

3. Monitoring by Means of the AE Technique: Experimental Results

3.1. Structural Stability Assessment of In-Service Double Girder Crane

We report the AE analysis of a steel double girder bridge crane with a service load of $2x10^3$ daN and over 30 years of service. An array of eight piezoelectric sensors was fixed on one of the two beams of the bridge crane: a pair of sensors for each of the two ends, in correspondence of the supports, and the other four were positioned in correspondence of the middle part of the monitored element (Figure 6).

The AE sensors were connected to an 8-channel acquisition system AEmission® which implements algorithms for automatic analysis of AE signal parameters, i.e., arrival time, duration, amplitude and counts number (total number of signal detection threshold crossings). The signal amplitude detection threshold was set to 10 mV in order to exclude from the analysis structural vibrations generated while the girder crane was moving in its unloaded conditions. Then, a 120-hour monitoring period started under the usual loading conditions. In order to suppress possible voltage spikes, acquired AE signals with duration <3 μs and counts number < 3 were filtered out.

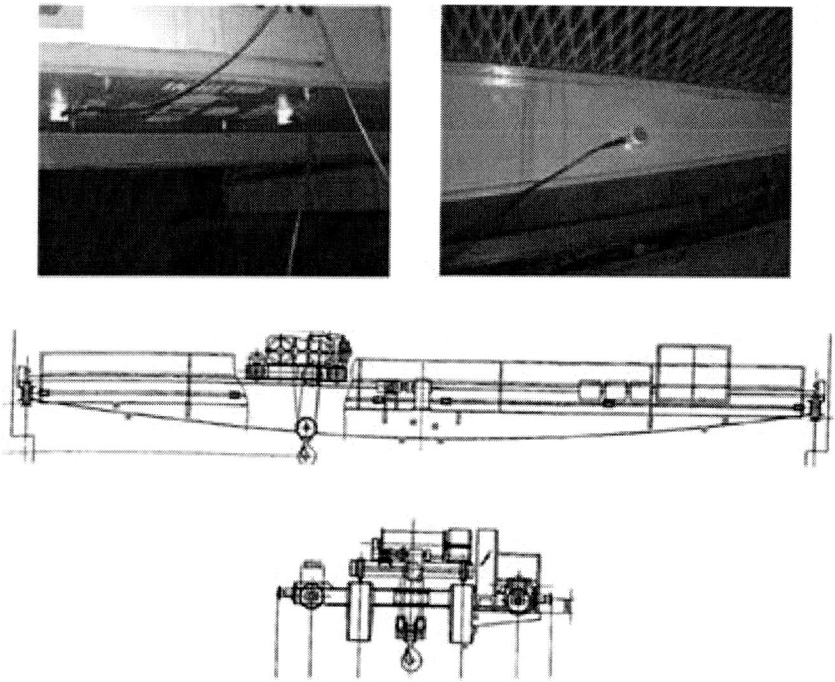

Figure 6. Pictures of the girder bridge crane with applied sensors position.

Zones undergoing structural damage were identified from the filtered signals exploiting different inversion methods, e.g., spatial location of active and propagating cracks by triangulation equations 32. Damage assessment was carried out exploiting b-value analysis.

From the analysis of the AE time series it is possible to correlate the AE activity with the most intense activity of the crane. The variability of the *b*-value (between 3.8 and 2.8) was far from critical ($b = 1.5$) or collapse ($b = 1$) condition (Figure 7). By means of the localization procedure, triangulation of the AE signals was carried out to identify the AE sources. The localization results show an asymmetric distribution of the AE sources, localized prevalently towards a semi-beam of the crane (Figure 8).

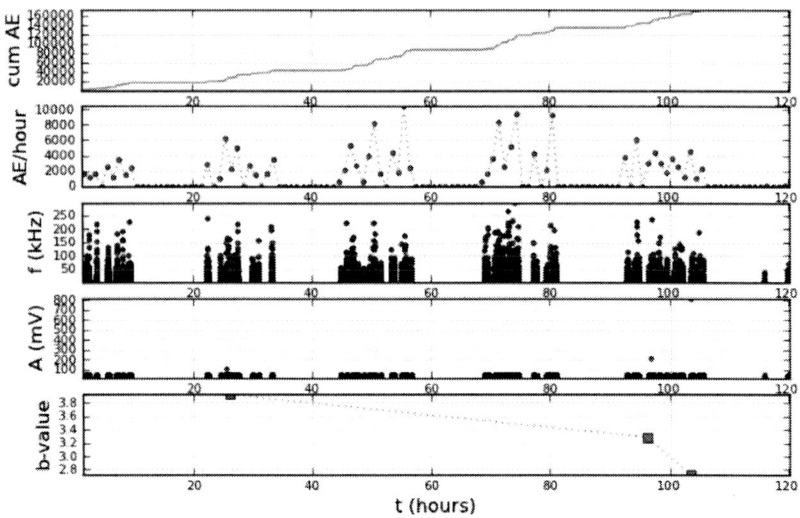

Figure 7. AE time series of events, frequencies, amplitudes and b-values (from top to bottom).

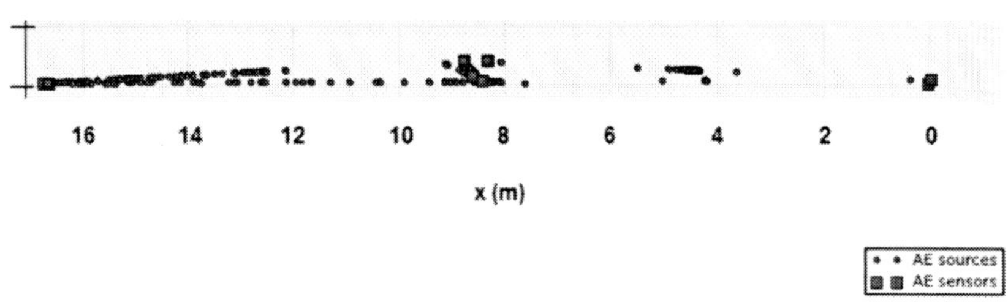

Figure 8. Output of the software showing the localized AE sources (dots) and transducers positions (squares).

3.2. Structural Stability Assessment of a R.C. Beam Tested in Bending LSM

By means of AE technique, we monitored and analyzed a carbon fiber reinforced concrete beam with nonrectangular cross-section tested in bending (Figure 9). The examined beam sustains the roof of a school rising in an urban area close to Turin, in the northern part of Italy. This single-story building was erected in 1974, covering a rectangular area of 24x30 m^2. The beam, with a reversed V cross-section (Figure 10), was affected by microcracking due to overloading. To improve the loading carrying capacity and reduce the effects of microcracking, the beam was externally reinforced with FRP sheets after a preliminary complete removal of the existing overload.

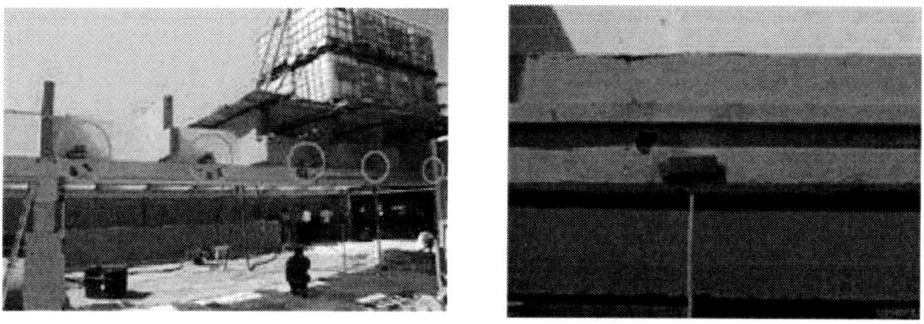

Figure 9. AE Monitoring of R.C: beam tested in bending (left); AE sensor (right).

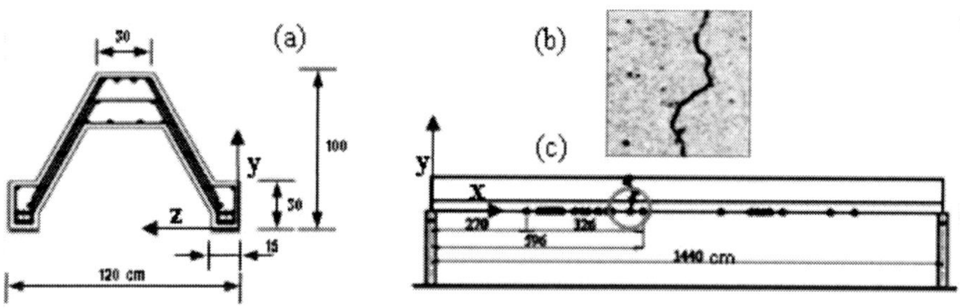

Figure 10. Scheme of the beam cross-section (a); photo of flexural crack between sensors 3 and 4 (using an optical microscope with 100X magnification) (b); scheme of the beam indicating localized AE sources (c).

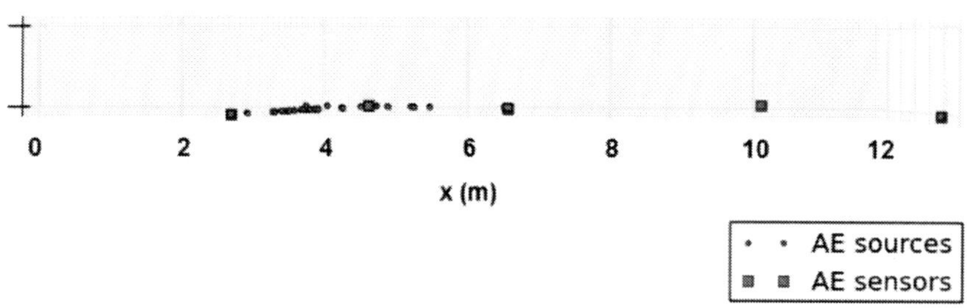

Figure 11. Output of the software showing the localized AE sources (dots) and transducers positions (squares).

In order to evaluate the loading carrying capacity of the retrofitted beams, an *in-situ* loading test of the retrofitted beam was carried out. The load was applied by filling 16 vessels (each one with a capacity of 1000 l) with water, equivalent to a total vertical load of 160 kN, symmetric with respect to the center of the beam (Figure 9). Gradually filling the vessels, four loading steps of 40 kN per step were applied.

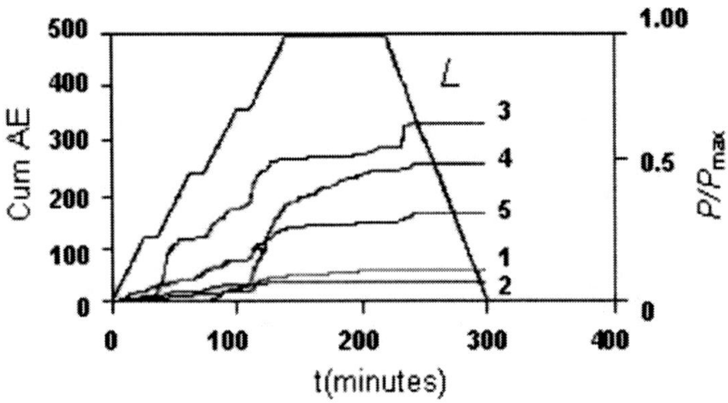

Figure 12. AE during the loading test: sensors 1 and 2 were fixed at the beam ends, sensors 3 and 4 in the middle of the beam and sensor 5 in an intermediate position. The loading history is reported as well.

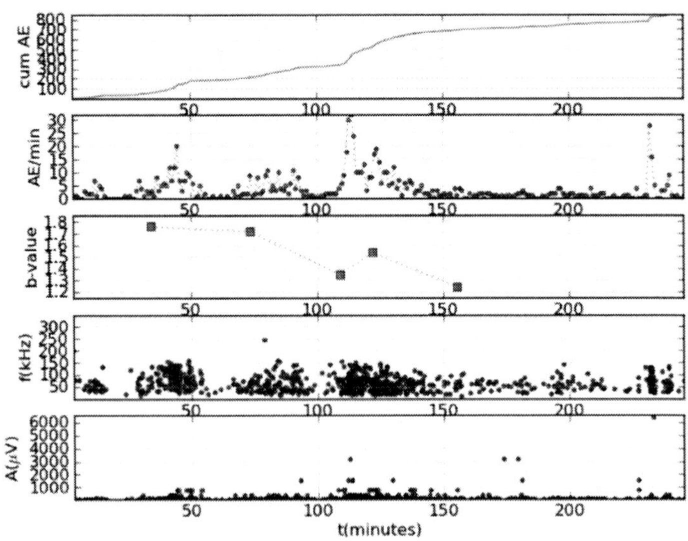

Figure 13. AE time series of events, b-values, frequencies and amplitudes (from top to bottom).

After each step, the applied load was kept constant for 10 minutes. The whole loading test last approximately 3 hours. A monitored mid-span deflection of 11 mm was observed at the end of the loading test. The full flexural capacity of the FRP-sheets was not reached at the end of the test, since a macroscopic delamination was not been seen. On the other hand, flexural cracks were observed in the middle of the beam and at 1/3 of the total span from the right support (Figure 10).

AE sources location along the beam was put forward using an array of five AE transducers, applied to the lateral surfaces of the beam. Applying the triangulation equations, the AE sources were determined and depicted in Figure 78 with black points and in Figure 11 with red points. In the considered loading range, micro-slips between FRP and concrete were not high enough to produce delamination. In fact, AE transducers detected the onset of debonding only. Indeed, delamination of the FRP sheets together with a thin layer of concrete takes place only when shear cracks develop close to the support. The progress of the acoustic events was then in agreement with this cracking progression.

Moreover, the time evolution of the AE counting numbers detected by the AE transducers is shown in Figure 80. Transducers 3 and 4 were close to flexural cracks and started to detect acoustic emissions from the beginning of the loading test. At the end, they detected the highest number of acoustic emissions, followed by transducers 1 and 2, close to the beam support. It is important to notice that acoustic emissions were not detected by the transducers 1 and 2 for vertical loads lower than 80 kN. This result is in agreement with the typical progression of cracking and collapse for retrofitted beams.

The results of the *b*-value analysis were in good agreement with the previous observations. In fact, AE transducers detected only the onset of debonding, with a *b*-value comprised between 1.2 and 1.7 (middle diagram of Figure 13). This means that the structure was still in stable conditions during the loading test, i.e., in a stage where macrocracks only start to appear [18].

CONCLUSION

By means of two application examples of AE monitoring presented ahead, we have illustrated the possibility of covering large structural elements and systems requiring continuous monitoring. In particular, we have shown:

- identification of active damage and defects in the structure through the cumulative count of all AE signals versus time;
- real time monitoring and assessment of damage evolution and structural stability through the b-value analysis;
- 3-dimensional location of active defects and propagating cracks.

REFERENCES

[1] Richter, C.F. 1958. *Elementary Seismology*. San Francisco and London: W.H. Freeman and Company.

[2] Omori, F. 1894. "On aftershocks" *Rep. Imp. Earthquake Invest. Comm.* 2: 103-109.

[3] Kanamori, H. and Anderson, D.L. 1975. "Theoretical basis of some empirical relationships in seismology". *Bulletin of the Seismological Society of America* 65:1073-1095.

[4] Aki, K. 1981. "A probabilistic synthesis of precursory phenomena". In *Earthquake Prediction*, edited by D.W. Simpson and P.G. Richards, 566-574. American Geophysical Union.

[5] Mogi, K. 1962. "Study of elastic shocks caused by the fracture of heterogeneous materials and its relation to earthquake phenomena". *Bulletin of Earthquake Research Institute* 40:125-173.

[6] Turcotte, D. L. 1997. *Fractal and Chaos in Geology and Geophysics*. Cambridge: Cambridge University Press.

[7] Bonnet, E., Bour, O., Odling, N. E., Davy, P., Main, I. P., Cowie, P. and Berkowitz, B. 2001. "Scaling of fracture systems in geological media". *Reviews of Geophysics* 39: 347-383.

[8] Bak, P., Christensen, K., Danon, L. and Scanlon, T. 2002 "Unified scaling law for earthquakes". *Physics Rev. Lett.* 88:178501/1-4.

[9] Corral, A. 2006. "Statistical Features of Earthquake Temporal Occurrence". In Modeling Critical and Catastrophic Phenomena in Geoscience, edited by P Bhattacharya et al, 191-221. Berlin: *Springer Lecture Notes in Physics*, 705.

[10] Davidsen, J., Stanchits, S. and Dresen, G. 2007. "Scaling and universality in rock fracture". *Physics Rev. Lett.* 98:125502/1-4.

[11] Kun, F., Carmona, H. A., Andrade, J. S. Jr., and Herrmann, H. J. 2008. "Universality behind Basquin's Law of Fatigue". *Physics Rev. Lett.* 100:094301/1-4.

[12] Tosi, P., De Rubeis, V., Loreto, V. and Pietronero, L. 2004. “Space-time combined correlation integral and earthquake interactions”. *Annals of Geophysics* 47:1849-1854.
[13] Geiger, L. 1910. “Probability method for determination of earthquakes epicenters from the arrival time only”. *Nachrichten von der Koniglichen Gesellschaft der Wissenschaften zu Gottingen* 4:331-349.
[14] Aki, K. and Richards, P.G. 1980. *Quantitative Seismology: Theory and Methods*. New York: W.H. Freeman and Company.
[15] Fletcher, R. 1987. *Practical Methods of Optimization*. New York: J. Wiley and Sons.
[16] Lee, W. H. K. and Lahr, J. C. 1972. “HYPO71: A computer program for determining hypocenter, magnitude, and first motion pattern of local earthquakes”. *Open File Report*, U.S: Geological Survey.
[17] Carpinteri, A., G. Lacidogna, G., A. Manuello, A. And G. Niccolini, G. 2016. “A study on the structural stability of the Asinelli Tower in Bologna”. *Structural Control and Health Monitoring* 23:659-667.
[18] Colombo, S., Main, I. G. and Forde, M. C. 2003. “Assessing damage of reinforced concrete beam using “b-value” analysis of acoustic emission signals”. *Journal of Materials in Civil Engineering ASCE* 15:280-286.
[19] Shiotani, T., Fujii, K., Aoki, T. and Amou, K. 1994. “Evaluation of progressive failure using AE sources and improved b-value on slope model tests”. *Prog. Acoust. Emission VII*: 529-534.
[20] Carpinteri, A., Lacidogna, G., Niccolini, G. and Puzzi, S. 2008. “Critical defect size distributions in concrete structures detected by the acoustic emission technique”. *Meccanica* 43:349-363.
[21] Carpinteri, A., Lacidogna, G.; Corrado, M. and Di Battista, E. 2016. “Cracking and crackling in concrete-like materials: A dynamic energy balance” *Engineering Fracture Mechanics*, 155, 130-144.
[22] Lacidogna, G., Manuello, A., Niccolini, G. and Carpinteri, A. 2015. “Acoustic emission monitoring of Italian historical buildings and the case study of the Athena temple in Syracuse”. *Architectural Science Review*, 58:290-299.

EDITORS' CONTACT INFORMATION

Dr. Claudia Barile
Dipartimento di Meccanica, Matematica e Management (DMMM)
Politecnico di Bari
Viale Japigia 182 – 70126
Bari, Italy
claudia.barile@poliba.it

Dr. Giovanni Pappalettera
Dipartimento di Meccanica, Matematica e Management (DMMM)
Politecnico di Bari
Viale Japigia 182 – 70126
Bari, Italy
giovanni.pappalettera@poliba.it

INDEX

A

acoustic signature, 1, 3, 4, 24, 30, 37, 51
adhesion, 119, 130
AE signal energy, 7, 37
aerospace, 26, 75
aggregation, 18
AISI 430F/1, 53, 54, 55, 66, 69
algorithm, 3, 12, 13, 15, 16, 17, 18, 28, 30, 31, 32, 33, 35, 37, 45, 46, 47, 48, 56, 140
alters, 35
amplitude, 3, 8, 9, 18, 19, 22, 25, 26, 31, 32, 33, 36, 39, 40, 54, 55, 63, 64, 65, 66, 67, 69, 115, 117, 118, 121, 123, 126, 141, 142
anisotropy, 42, 136
annealing, 70
Artificial Neural Networks, 48
assessment, 22, 44, 116, 127, 134, 141, 143, 148
attribution, 19
automotive applications, 129
aviation industry, 71

B

background noise, 58
beams, 40, 133, 142, 145, 147
bending, 110, 112, 133, 144, 145
benefits, vii
biodegradable polymers, 116
biopolymers, 116
boundary surface, 3
broadband, 50, 77
by-products, 114

C

calibration, 38, 47, 50, 77
carbon, 26, 43, 44, 49, 55, 71, 76, 93, 96, 110, 111, 112, 115, 126, 133, 144
case studies, 134, 149
cellulose, 114, 128
ceramic, 26, 36, 37, 38, 39, 40, 44, 45, 46, 47, 49, 50, 55
ceramic matrix composites (CAM), 26, 28, 38, 39, 40, 44, 46, 49, 50
chemical, 27, 114, 115, 127
classification, 2, 4, 12, 15, 16, 21, 25, 26, 28, 29, 30, 31, 32, 34, 37, 40, 43, 47, 48, 50, 74, 109, 116, 117, 122
cluster analysis, 4, 12, 13, 48
clustering, 1, 3, 4, 5, 8, 10, 12, 13, 15, 24, 26, 29, 30, 31, 32, 42, 45, 46, 47, 48, 74, 96, 101, 103, 104, 105, 106, 112
clustering method, 1, 4, 12, 47, 74, 96, 104, 112
CMC, 6, 26, 47

community, vii
complexity, 126
compliance, 78
composite, viii, 1, 3, 6, 7, 22, 24, 26, 36, 30, 37, 38, 39, 40, 42, 43, 44, 45, 46, 47, 48, 49, 50, 51, 55, 71, 73, 74, 75, 76, 78, 82, 85, 86, 93, 105, 108, 109, 110, 111, 112, 113, 114, 115, 117, 118, 119, 120, 121, 123, 124, 125, 126, 127, 128, 129, 130, 131
composition, 4
compression, 19, 122
computation, 15, 82, 140
conditioning, 140
conductivity, 55
configuration, 21, 38, 39, 40, 69, 81, 83
constant load, 76
construction, 29
consumption, 115
contamination, 35
convergence, 15, 140, 141
cooling, 21, 23, 49
copolymers, 130
correlation, 3, 10, 11, 30, 44, 126, 149
correlation coefficient, 11
corrosion, 3, 27, 29, 44, 46, 47, 49
cost, 134, 140
covering, 134, 144, 147
crack location, 134
cracks, 2, 22, 27, 42, 125, 126, 143, 147, 148
crevice corrosion, 27, 46, 49
critical stress intensity factor, 65
cross-validation, 16
cumulative distribution function, 38
cycles, 18, 19, 20, 54, 61, 62, 63, 66, 67, 68, 69, 125, 128

D

damage assessment, 44, 134, 141, 143
damage characterization., 74, 82
damage identification, 2
damage localisation, 126
damage monitoring, 1, 47, 50, 71
damping, 115, 128
data analysis, 3, 65, 67
data set, 11, 13, 14, 15, 28, 30, 31, 32, 33, 34, 105
data structure, 31, 32, 43
database, 28
Davies-Bouldin validity index, 13
decision trees, 18
decomposition, 46, 100, 101
defects, 53, 73, 113, 134, 148
deformation, vii, 24, 115, 122, 126, 128, 134
degradation, 133
delamination, 73, 74, 75, 78, 79, 81, 82, 83, 85, 86, 87, 88, 90, 92, 93, 94, 96, 97, 98, 109, 110, 111, 112, 116, 125, 127, 147
dendrites, 21
dendrogram, 11
denoising, 8
depth, 8
derivatives, 69, 140
descriptors, 2, 3, 4, 6, 8, 9, 10, 11, 12, 26, 30, 31, 32, 36, 38, 39
designers, 129
detection, vii, 24, 42, 49, 54, 64, 69, 78, 126, 135, 142
deviation, 78
diffusion, 42
discriminant analysis, 35
discrimination, 2, 3, 5, 26
dislocation motion, vii, 18, 19
disorder, 21
dispersion, 30, 36, 50
displacement, 30, 40, 76, 78, 79, 82
distribution, 14, 19, 21, 38, 39, 63, 64, 65, 79, 81, 105, 117, 124, 125, 126, 127, 141, 143
durability, 1, 113, 130

E

earthquakes, vii, 141, 148, 149
elastic deformation, 82
electrical resistance, 44
electron, 74, 109
energy, 2, 4, 5, 7, 8, 18, 19, 20, 21, 27, 31, 32, 37, 40, 46, 50, 55, 74, 75, 77, 79, 80, 81, 82, 85, 88, 93, 94, 96, 97, 98, 103, 120, 121, 123, 127, 134, 149
engineering, vii, 54, 73, 75, 114, 115
enlargement, 29
environmental degradation, 119, 121, 127
environmental protection, 113
evolution, 2, 6, 7, 22, 24, 28, 33, 35, 42, 49, 54, 55, 61, 64, 66, 67, 69, 71, 74, 75, 126, 133, 141, 147, 148
experimental condition, 105
expertise, viii, 54
extraction, 5, 8

F

falling weight impact, 124
fatigue damage propagation, 53, 54
fatigue tests, 18, 26, 38, 45, 46, 56, 69, 127
feature selection, 5
FFT, 87, 89, 96, 99, 122
fiber, viii, 1, 2, 22, 24, 26, 30, 40, 42, 43, 44, 47, 49, 55, 71, 75, 85, 86, 87, 88, 89, 92, 94, 96, 109, 110, 112, 127, 128, 129, 130, 131, 133, 144
fiber failure, 24, 26
fibrillation, 118, 127
filament, 24
filters, 54, 102
finite element method (FEM), 40, 56
fitness, 15
flax fiber, 128, 131
force, 76
formation, 18, 20
fracture precursors, 43, 134
fracture processes, 134
fracture toughness, 74, 75, 78, 110
fractures, 126, 134
fragments, 25
France, 1
frequency distribution, 75, 116
friction, 115, 121
FRP, 145, 147

G

gel, 56
geometry, 2, 40, 42, 56, 58
glycerol, 115
graph, 54, 63
growth, 19, 22, 49, 71, 72, 74, 85, 86, 87, 88, 90, 92, 93, 94, 96, 97, 98, 110, 111, 112, 114

H

healing, 26
health, 2, 45, 47
hemp, 116, 119, 120, 122, 125, 126, 128, 129, 130, 131
heterogeneity, 15
history, 40, 88, 146
humidity, 122
hybrid, 3, 49, 69, 129
hypothesis, 26

I

identification, 1, 2, 3, 4, 19, 24, 27, 30, 36, 37, 41, 43, 45, 46, 47, 48, 51, 69, 79, 110, 119, 148
image, 36, 37, 58, 60, 61, 106, 108
impact energy, 125
impregnation, 114, 120, 121
incidence, 73
India, 112
industries, vii, 115
inequality, 138

initiation, 2, 7, 18, 22, 24, 69, 74, 75, 77, 78, 79, 82, 83, 109, 110
inspections, 54
Instron, 56
integration, 84
integrity, vii, 2, 133
interface, 22, 75, 115, 119, 120, 121, 122
interfacial shear strength, 120, 121, 127
interpretability, 48
inversion, 133, 134, 143
Iran, 73
Italy, viii, 53, 113, 133, 144
iteration, 139, 140, 141

K

kinetics, 2, 37, 42
k-means algorithm, 12, 15, 31, 32, 45
k-nearest neighbor, 1, 16, 26, 48
Kohonen map, 1, 16

L

lactic acid, 116, 126, 129, 131
laminated composites, viii, 73, 74, 75, 78, 85, 86, 109, 111
large-sized structures, 133, 134
learning, 16, 17, 18, 28, 29, 34, 50
learning library, 29
life cycle, 115
life expectancy, 133
lifetime, 1, 2, 17, 46, 49, 50, 73
localization, 21, 55, 56, 57, 69, 74, 85, 86, 87, 133, 136, 143
localized corrosion, 3, 27, 47
lumen, 129

M

machinery, 134
magnitude, 3, 134, 141, 149
majority, 12, 16, 28, 29, 32
manufacturing, 21, 73, 114
martensitic transformation, 1, 2, 18, 19, 21, 43, 48
mass, 21
material ageing, 24
material degradation, 121
material surface, 7, 38
materials, vii, 1, 2, 4, 21, 24, 26, 42, 45, 48, 49, 54, 71, 73, 74, 75, 78, 85, 86, 105, 110, 111, 112, 113, 114, 116, 117, 118, 124, 126, 127, 129, 130, 134, 148, 149
matrix, 3, 10, 22, 24, 25, 26, 36, 37, 38, 39, 40, 42, 44, 46, 49, 50, 55, 58, 71, 75, 85, 94, 99, 103, 104, 105, 106, 109, 110, 114, 115, 119, 120, 121, 122, 126, 127, 130, 139, 140, 141
measurement, 13, 30, 44, 49, 126, 127, 134
mechanical properties, 26, 44, 114, 129, 130
mechanical testing, 117
media, 148
median, 56
membership, 16, 104, 105
memory, 48
metallic specimens, 54
metals, vii, viii, 43, 50, 55, 136
methodology, 12, 29, 47, 86, 88
microscope, 145
mislabeled data, 34
modal acoustic emission, 40, 51
modelling, viii, 2, 4, 30, 40, 43, 49
models, 42, 50
modulus, 6, 20, 49, 136
multivariate analysis, 122

N

NaCl, 28
natural selection, 15
NDI, 46
neural network, 3, 16, 23, 46, 48
neurons, 16, 17
New Zealand, 117
nodes, 8

non-linear equations, 137
nucleation, 56

O

oil, 120, 130
operations, 15
original training, 18, 35
oxygen, 49

P

PAN, 26
partition, 12, 13, 33, 105
pattern recognition, 2, 5, 12, 15, 18, 23, 45, 46, 47, 48, 50, 112
PCA, 1, 11, 26
permit, 104
pH, 28
phase transformation, 2, 21
physical phenomena, 104
pitch, 44
plant fibre composites, 113, 115, 117, 118, 119, 120, 125, 126
plants, 114
plastic deformation, vii, 2
plasticity, 18, 19
plastics, 127
polymer, 24, 44, 45, 47, 110, 116, 120, 128, 129
polymer matrix, 44, 110, 128
polypropylene, 44, 55, 110, 120, 122, 126, 130, 131
polyurethane foam, 43
population, 15
power law, 19, 21, 43
power relations, 94
principles, 43
prognosis, 133
propagation, 2, 5, 7, 8, 18, 22, 30, 36, 37, 39, 40, 53, 54, 56, 60, 61, 67, 69, 70, 72, 74, 75, 79, 85, 93, 96, 109, 126, 127, 136

Q

quality control, 117
quantitative AE, 4, 43

R

radiography, 110, 134
radius, 17, 56
random forests, 17, 18, 28, 47, 48
real time, 2, 4, 113, 148
real-time monitoring, 134
reciprocity, 50
recognition, 8, 35, 45, 46, 50
refractories, 1, 21, 49
reinforcement, 29, 116, 117, 120, 121, 131
relaxation, 70
reliability, 1, 28, 78
researchers, vii, viii, 78
residuals, 138, 139, 140
resolution, 61, 69
response, 7, 41, 111, 131
room temperature, 1, 18, 24

S

scaling, 43, 102, 134, 141, 148
scaling law, 134, 148
scanning electron microscopy, 127
scatter plot, 65
scattering, 86, 88
school, 144
science, 110
security, 16, 28
seed, 120
self-organization, 16
self-similarity, 134
sensitivity, 2, 127
sensors, 3, 5, 6, 7, 27, 36, 37, 38, 39, 40, 41, 42, 47, 50, 53, 55, 56, 57, 58, 59, 65, 67, 69, 74, 77, 86, 110, 111, 127, 133, 134, 136, 137, 138, 139, 142, 143, 145, 146

service loads, 134
set theory, 104
shear strength, 120, 121, 127
signals, viii, 2, 3, 4, 5, 6, 7, 8, 9, 12, 13, 14, 15, 16, 18, 19, 21, 22, 23, 24, 25, 26, 27, 29, 30, 31, 32, 34, 36, 37, 38, 39, 40, 41, 42, 45, 46, 47, 48, 56, 74, 75, 77, 79, 82, 83, 85, 86, 87, 88, 89, 96, 101, 105, 109, 116, 117, 122, 124, 126, 133, 141, 143, 148, 149
silane, 121, 123
silhouette, 13, 14, 15, 31, 32, 37
silicon, 56, 77
simulation, 40
sinusoidal loading, 26, 56
sodium hydroxide, 114, 115
solution, 2, 14, 15, 21, 28, 31, 32, 137, 138, 139, 141
solution space, 15
spatial location, 143
specialization, viii
species, 114, 119, 120
specifications, 76
stability, 148, 149
standard deviation, 8
standardization, 8, 38
starch, 126, 128, 131
state, 7, 69
statistics, 43
steel, 18, 28, 30, 43, 48, 49, 55, 65, 69, 71, 72, 142
stress, vii, 3, 27, 49, 53, 54, 55, 56, 70, 75, 85, 118, 134
stress intensity factor, 56
structure, vii, 4, 12, 21, 31, 32, 54, 55, 74, 75, 113, 121, 147, 148
supervised classification, 4, 15, 26, 28, 29, 34, 37, 47, 50
surface energy, 55
surface treatment, 121
sustainability, 113, 114
synthesis, 148

T

tar, 44
techniques, 23, 30, 44, 48, 50, 53, 55, 69, 72, 104, 111, 127, 139, 141
temperature, 21, 23, 26, 28, 38, 43, 46, 47, 50, 54, 55, 58, 60, 61, 62, 63, 69, 76, 116, 122, 127, 131
tensile strength, 121
tension, 26, 47, 122, 131
testing, 2, 18, 26, 28, 51, 70, 113, 114, 115, 125, 134
textiles, 114, 126
thermal energy, 71
thermal expansion, 22
thermograms, 62
time series, 143, 144, 146
titanium, 71, 110
topology, 17
torsion, 115
total energy, 19, 103
training, 16, 17, 18, 24, 27, 35
transformation, 1, 2, 18, 19, 21, 30, 43, 48
transition temperature, 21
translation, 102
treatment, 114, 115, 121, 122, 127
trial, 88, 139
triangulation, 55, 65, 136, 137, 138, 143, 147
two-dimensional space, 33

U

ultrasound, 53, 134
uniform, 3, 76
universality, 21, 148
unsupervised classification, 2, 12, 26, 31, 32

V

vacuum, 38
validation, 21, 48

variables, 104
vector, 4, 8, 13, 16, 17, 104, 140
vegetable fibres, 113, 114, 127
velocity, 6, 7, 47, 74, 85, 86, 131, 136, 137, 138
vibration, 115, 128
viscosity, 38

W

water, 24, 42, 115, 121, 127, 130, 146
wave propagation, 7, 37, 41, 42, 49, 136, 138
wavelet, 8, 45, 46, 74, 100, 102, 112
wood, 120

X

XRD, 22

Z

zirconia, 21, 49